AF360701

Atmospheric Heavy Metal and Nitrogen Deposition Using Mosses as Biomonitors

Atmospheric Heavy Metal and Nitrogen Deposition Using Mosses as Biomonitors

Editor

Antoaneta Ene

MDPI • Basel • Beijing • Wuhan • Barcelona • Belgrade • Manchester • Tokyo • Cluj • Tianjin

Editor
Antoaneta Ene
Chemistry, Physics and
Environment
Dunarea de Jos University of
Galati
Galati
Romania

Editorial Office
MDPI
St. Alban-Anlage 66
4052 Basel, Switzerland

This is a reprint of articles from the Special Issue published online in the open access journal *Atmosphere* (ISSN 2073-4433) (available at: www.mdpi.com/journal/atmosphere/special_issues/ Heavy_Metal_Deposition_Bioindicators).

For citation purposes, cite each article independently as indicated on the article page online and as indicated below:

LastName, A.A.; LastName, B.B.; LastName, C.C. Article Title. *Journal Name* **Year**, *Volume Number*, Page Range.

ISBN 978-3-0365-1572-4 (Hbk)
ISBN 978-3-0365-1571-7 (PDF)

Contents

About the Editor

Antoaneta Ene

Antoaneta Ene (Prof.dr.habil.eng.) obtained her Ph.D. in Particle Physics in 1997 at the University of Bucharest, Romania, studying the elemental composition of raw materials and finite products in iron and steel industry using atomic and nuclear methods. She has a MSc. in Environmental Sciences (2007) and obtained a Habilitation in Industrial Engineering (2015) at Dunarea de Jos University of Galati, Romania.

Her research interests include the determination of trace elements and radioisotopes in industrial, geological, food and environmental materials; heavy metal atmospheric deposition using mosses as biomonitors; identification of microplastics in personal care products; analysis of composition and structure of advanced materials using nuclear techniques and imaging microscopy; monitoring of toxic substances in the environment, including pharmaceutical residues and metabolites, and assessment of their impact upon ecosystems and human health.

Preface to "Atmospheric Heavy Metal and Nitrogen Deposition Using Mosses as Biomonitors"

Air pollution has a negative impact on various compartments of ecosystems, posing a threat to the natural environment and human health, and also causing significant economic damage.

Due to their specific features, mosses are recognized as one of the main bioindicators and biomonitors of air contamination, with toxic elements including those originating from anthropogenic and natural sources. The determination of elemental concentrations in mosses is easier and cheaper than conventional precipitation analysis, and a much higher sampling density can be achieved by employing moss biomonitoring.

In recent decades, naturally growing mosses have been used successfully in biomonitoring campaigns for checking the atmospheric fallout of heavy metals and nitrogen (N) across Europe, and the approach has been extended in many regions of the world for characterizing multi-elemental deposition sources.

Quantification of heavy metals and N in selected moss species provides a time-integrated measure of the spatial patterns and temporal trends of heavy metal deposition from the atmosphere to terrestrial ecosystems, and a good indication of ecosystems at risk from high N deposition.

The Special Issue *"Atmospheric Heavy Metal and Nitrogen Deposition Using Mosses as Biomonitors"* belongs to the section *Air Quality and Human Health* of the journal *"Atmosphere"* and includes a collection of papers related to aspects of passive moss biomonitoring of air quality in various regions of the world regarding the pollution sources of potentially toxic elements, heavy metal air pollution in the lockdown period due to the COVID-19 pandemic, trends in element atmospheric deposition and relevance for ecological integrity and human health. Most of the studies were carried out in the framework of the *International Cooperative Program on Effects of Air Pollution on Natural Vegetation and Crops* (ICP Vegetation) of the *United Nations Economic Commission for Europe* (UNECE).

Antoaneta Ene
Editor

 *atmosphere*

Article

Atmospheric Heavy Metal Deposition in North Macedonia from 2002 to 2010 Studied by Moss Biomonitoring Technique

Lambe Barandovski [1,2], **Trajče Stafilov** [3,*] , **Robert Šajn** [4], **Marina Frontasyeva** [2] **and Katerina Bačeva Andonovska** [5]

[1]   Faculty of Natural Sciences and Mathematics, Institute of Physics, Ss. Cyril and Methodius University, POB 162, 1000 Skopje, Macedonia; lambe@pmf.ukim.mk

[2]   Frank Laboratory of Neutron Physics, Joint Institute for Nuclear Research, Moscow Region, 141980 Dubna, Russia; marina@nf.jinr.ru

[3]   Faculty of Natural Sciences and Mathematics, Institute of Chemistry, Ss Cyril and Methodius University, POB 162, 1000 Skopje, North Macedonia

[4]   Geological Survey of Slovenia, Dimičeva 14, 1000 Ljubljana, Slovenia; Robert.Sajn@GEO-ZS.SI

[5]   Research Center for Environment and Materials, Academy of Sciences and Arts of the Republic of North Macedonia MANU, Krste Misirkov 2, 1000 Skopje, North Macedonia; kbaceva@manu.edu.mk

*   Correspondence: trajcest@pmf.ukim.mk; Tel.: +3-897-035-0756

Received: 27 July 2020; Accepted: 27 August 2020; Published: 30 August 2020

Abstract: Moss biomonitoring technique was used for a heavy-metal pollution study in Macedonia in the framework of the International Cooperative Program on Effects of Air Pollution on Natural Vegetation and Crops (UNECE IPC Vegetation). Moss samples ($n = 72$) were collected during the summers of 2002, 2005, and 2010. The contents of 41 elements were determined by neutron activation analysis, atomic absorption spectrometry, and inductively coupled plasma atomic emission spectrometry. Using factor and cluster analyses, three geogenic factors were determined (Factor 1, including Al, As, Co, Cs, Fe, Hf, Na, Rb, Sc, Ta, Th, Ti, U, V, Zr, and rare-earth elements–RE; Factor 4 with Ba, K, and Sr; and Factor 5 with Br and I), one anthropogenic factor (Factor 2, including Cd, Pb, Sb, and Zn), and one geogenic-anthropogenic factor (Factor 3, including Cr and Ni). The highest anthropogenic impact of heavy metal to the air pollution in the country was from the ferronickel smelter near Kavadraci (Ni and Cr), the lead and zinc mines in the vicinity of Makedonska Kamenica, Probištip, and Kriva Palanka in the eastern part of the country (Cd, Pb, and Zn), and the former lead and zinc smelter plant in Veles. Beside the anthropogenic influences, the lithology and the composition of the soil also play an important role in the distribution of the elements.

Keywords: air pollution; moss; potentially toxic metals; Macedonia; biomonitoring

---

1. Introduction

Elements and compounds that are released into the environment and damage the biosphere (soil, water, air) are treated as pollutants—regardless of whether they come from natural processes on the Earth's surface or human activities [1–3]. Atmospheric emissions, due to their widespread dispersion and exposure to living organisms, pose a major risk to human health [4,5]. The representation of pollutants is mostly dependent on the source of emissions, the amount of the released substances, their composition, state of matter, altitude, and weather conditions season, amount of atmospheric sediments, air currents, precipitation, acidity, temperature, humidity; the rate of deposition depends on several factors, such as meteorology wind velocity, relative humidity, and temperature; particle shape and size, and the chemical forms of the elements [1,2,6,7].

The monitoring of potentially toxic air pollutants has a significant role in understanding the spatial and temporal distribution of these pollutants; it serves as a tool in planning to reduce their harmful effects. Biomonitoring is a method to obtain quantitative information about certain characteristics of the biosphere. Among the most often used biomonitors are mosses, liches, grasses, etc. [5,7–10].

For more than 50 years, terrestrial mosses have been successfully used to monitor and map the atmospheric deposition of trace metals in many parts of the world [9–21]. Mosses are characterized by a small change in the morphology during the growing season. Due to their adaptability, mosses are widespread under different environmental conditions [22,23].

Global sources of pollution of the environment include activities, such as exploitation of natural resources and their processing through adequate technological processes, as well as non-ecological waste management. In this respect, the area of Macedonia does not deviate from the global average. In former Yugoslavia, Macedonia was an important producer of metals, such as copper, steel, ferroalloys, lead, zinc, cadmium, nickel, silver, and gold. Long-term mining and mining-related activities have led to the pollution of the environment, as reported in several studies [2,3,22,24–46].

For the first time in Macedonia, a heavy metal air pollution study, covering all the area of the country, was made in 2002 using mosses as biomonitors. In 2005 and 2010, the study followed the framework of the United Nations Economic Commission for Europe International Co-operative Program on Effects of Air Pollution on Natural Vegetation and Crops with heavy metals in Europe (UNECE ICP Vegetation). The first results of the air pollution biomonitoring research with heavy metals have increased the interest in applying biomonitoring with mosses that are focused on individual critical regions in Macedonia [37]. Investigations were conducted in the area of the Bučim copper mine in the vicinity of the town of Radoviš [27,40], in the area of ferronickel smelter plant near the town of Kavadarci (Ni and Cr) [30,47]; in the eastern parts of the country near Pb-Zn mines and flotation plants "Toranica" [3,41], "Zletovo" [39,42], and "Sasa" [42,43]. The study was also conducted in the vicinity of the abandoned As-Sb-Tl mine "Allchar" on the Kožuf Mountain [35], as well as in the surroundings of the thermoelectric coal-fired power plants in Bitola and Kičevo [48–50].

The aims of the paper are to monitor the temporal trends of elements and groups of elements related to air pollution trends by comparing the results of the three moss surveys in 2002, 2005, and 2010 in Macedonia. This study verifies the pollution sources identified with the first national survey performed in 2002 to distinguish anthropogenic from natural sources, to identify the deposition patterns over the entire study area, and to make the data available so that it can be compared with results of further surveys.

2. Materials and Methods

2.1. Study Area

The Republic of Macedonia is situated in the central part of the Balkan Peninsula (Figure 1) with an area of 25,713 km^2 between latitudes 40° and 43° N and between longitudes 20° and 23° E. Geographically, the investigated area is clearly defined by a central valley formed along the Vardar River, which is surrounded by mountain ranges. The country boasts 16 mountains higher than 2000 m, although most of its area lies between 500 and 1000 m. The lowest point (44 m) indicates an area on the cross point of the Vardar River; the highest point (2764 m) is the peak of Korab mountain [51]. Most of the population (2.057 million), lives in urban areas, and more than a quarter live in the capital city of Skopje [52].

The Republic of North Macedonia has three climatic types, due to specific natural and geographical features. The northern parts of the country are characterized by the mildly continental climate, while the area along the valleys of the Vardar River typically has a Mediterranean climate. There is also a mountainous climate, which is dominant in the mountainous region of the country. All three types of climate that spread and partly intertwine through the country have an impact on the properties of the individual seasons between regions. Winters are milder in the eastern areas, and summers are hotter and drier in the western areas [53].

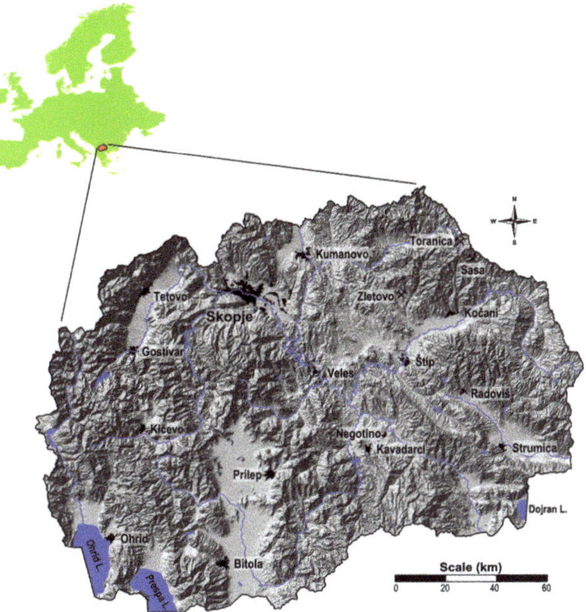

Figure 1. Map of the Republic of Macedonia.

The average annual precipitation varies from 500 to 600 mm in the central and eastern areas along the valley of the Vardar River to 1400 mm in the western mountainous area. The area with the least precipitation is between the Ovče Pole, Tikveš, and the Štip basins. Usually, the wind blows from the north/northwest toward the south/southeast or vice versa. However, the directions of the winds in various parts of Macedonia are heavily dependent on the orographic conditions. The average speed of the winds in the country is 6 m/s [53].

The country is enriched by reserves of several ores. The industrial sector represents 31.9 of Gross Domestic Product (GDP), while the agriculture sector represents 18.9%. The main industrial activities are situated in Skopje (steel and ferroalloys production and steel processing), Radoviš (copper mine and flotation), Veles (abandoned lead-zinc and cadmium smelter), Kavadarci (ferronickel smelter), Tetovo (abandoned ferrochrome smelter and present ferrosilicon smelter), and Kičevo and Bitola (thermoelectric power plants, where the lignite fuel is used). Three Pb-Zn mines and flotation plants are located (Sasa, Toranica, Zletovo) in the eastern part of the country; a previously active As-Tl-Sb mine "Allchar" is in the southern part of Macedonia. These active and abandoned industrial facilities have led to environmental pollution in the nearby and distant areas with different heavy metals, reported by several studies [24–45,47–50] and those studies realized recently [46,54–59].

Macedonia is constituted by several geotectonic units that were formed during different geological periods; these are generally spread out in NNW-SSE to N-S direction [44,59]. Contacts of units are marked by regional faults managed by the Laramide compression (Laramide orogeny phase) [44]. The country lies on four major tectonic units (Figure 2): The Serbo-Macedonian massif (SMM), the Vardar zone (VZ), the Pelagonian massif (PM), and the West-Macedonian zone (WMZ) [60,61]. The SMM and PM form the oldest complexes in which highly metamorphic Proterozoic rocks have developed; the area is characterized by widespread Riphean-Cambrian, Paleozoic, Mesozoic, and Cenozoic rocks [44]. The WMZ is lithologically composed of low-grade metamorphic rocks, and anchi-metamorphic

Paleozoic rocks and migmatites, Triassic and Jurassic sediments, and migmatites, and Tertiary sediments [44]. The VZ is a large and important lineament structure of the Balkan Peninsula [61]; it represents an area where a continuous thinning of the continental crust had occurred, and where, during the Lower Jurassic, it completely transited into the rift zone [44].

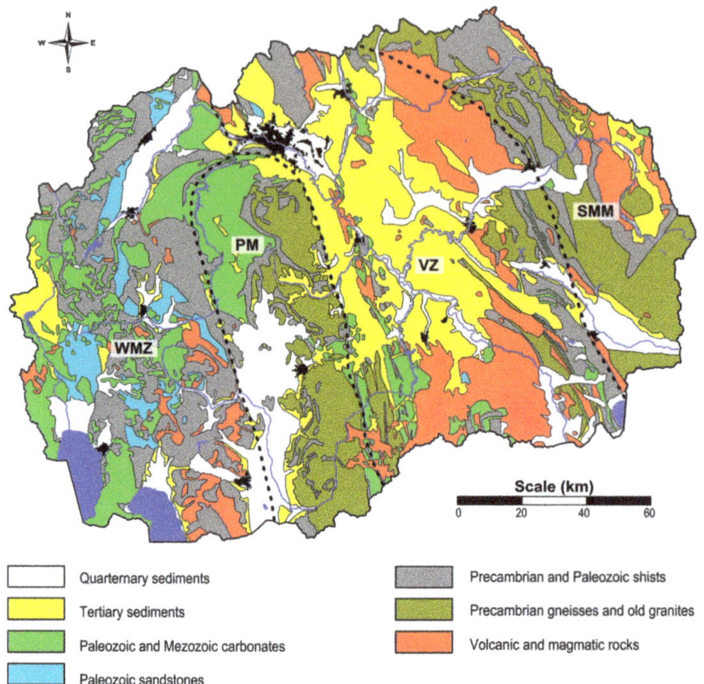

Figure 2. Lithological map of Macedonia (simplified according to a Geological map of SFR Yugoslavia, 1970). Dashed lines represent the areas of the major tectonic units: West-Macedonian zone (WMZ), Pelagonian massif (PM), Vardar zone (VZ), and Serbo-Macedonian massif (SMM).

2.2. Sampling and Sample Preparation

In this study, datasets from three different moss surveys were used for comparison among them. The first moss survey aiming to monitor heavy metal air pollution was carried out in 2002 in the entire territory of Macedonia. For this reason, samples from 72 sites of three moss species *Hypnum cupressiforme*, *Homalothecium lutescens*, and *Homalothecium sericeum* were collected from September to October [1,2]. The contents of 41 elements (Na, Mg, Al, Cl, K, Ca, Sc, Ti, V, Cr, Mn, Fe, Co, Ni, Cu, Zn, As, Se, Br, Rb, Sr, Zr, Mo, Ag, Cd, In, Sb, I, Cs, Ba, La, Ce, Sm, Eu, Tb, Hf, Ta, W, Au, Th, and U) in each sample were analyzed either by neutron activation analysis or by atomic absorption spectrometry. The next survey was conducted from August to September 2005; samples of *Hypnum cupressiforme* and *Homalothecium lutescens* were taken at the same 72 sites. Using neutron activation analysis and atomic absorption spectrometry, the content of 38 elements (Al, As, Ba, Br, Ca, Cd, Ce, Co, Cr, Cs, Cu, Dy, Eu, Fe, Hf, Hg, K, La, Mg, Mn, Mo, Na, Ni, Rb, Sb, Sc, Se, Sm, Sr, Ta, Tb, Th, U, V, W, Yb, Zn, and Zr) were determined in each sample [32]. The third moss survey was conducted in August and September 2010 at the same 72 locations, where the same moss species were collected as in the moss survey carried out in 2005. In addition to the techniques used in previous surveys, the atomic emission spectrometry with inductively coupled plasma (ICP-AES) was also used to determine 41 elements (Al, As, Au, Ba, Br, Ca, Cd, Ce, Cl, Co, Cr, Cs, Cu, Dy, Eu, Fe, Hf, Hg, I, K, La, Mg, Mn, Mo, Na, Ni, Pb, Rb, Sb, Sc, Se, Sm, Sr, Ta, Tb, Th, Ti, U, V, W, Zn, and Zr) in the samples [22,33].

To enable a comparison between individual surveys, the sampling locations mostly overlap (Figure 3), and the same biotope conditions were used. The sampling in 2002 followed the principles of European moss surveys [14,62]. In 2005, sampling was performed according to the guidelines of the ICP Vegetation Program—monitoring manual for the 2005/2006 survey, and the procedure used in the previous European moss surveys [63]. In 2010, the sampling followed the principles of the Convention on Long-Range Transboundary Ari Pollution (CLRTAP) and the International Cooperative Program on Effects of Air Pollution on Natural Vegetation and Crops monitoring manual for 2010/2011 survey and the procedure used in the previous European moss surveys [64].

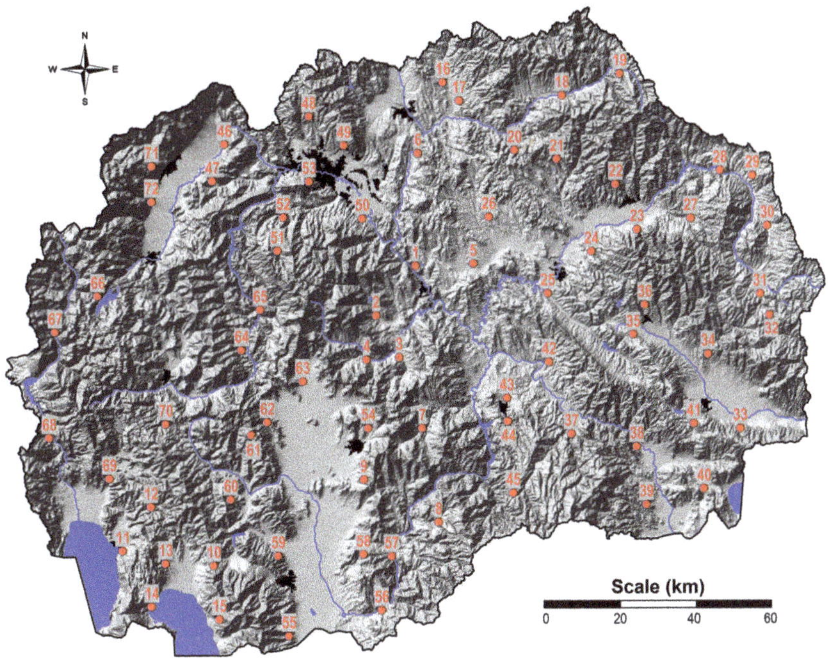

Figure 3. Location of sampling points.

In each sampling year, the samples were collected at least 300 m from a major road and populated area, and at least 100 m from small roads and single houses. Each sample was a mix of five to ten sub-samples taken from 50 m × 50 m. Sampling and sample handling was performed using separate polyethylene gloves for each sample, to prevent any contamination of the samples. Samples were properly marked and stored in paper bags. In cases where more than one moss species was collected at the same sampling site, the interspecies comparison showed no essential differences within error estimates.

Samples were brought to the laboratory where they were cleaned from extraneous material (litter and dead leaves) and dried to a constant weight for 48 h at 30–40 °C. Green-brown parts of the moss shoots represented the last three years of the moss growth and were selected to be subjected to the analysis of pollutant content. For neutron activation analysis (NAA), atomic absorption spectrometry (AAS), and atomic emission spectrometry with inductively coupled plasma (ICP-AES), the moss samples were prepared according to the description from previous studies, where quality control data are also given [2,22,32].

2.3. Statistical Methods

Data on chemical analyses obtained from previous studies [1,2,22,32,33] are used in the proposed report. To compare the results between different datasets, and simultaneously perform parametric and non-parametric statistical methods, the statistical software Statistica 13 (StatSoft, Inc., Tulsa, OK, USA) was used [65,66].

Since the datasets were non-normally distributed, a Box-Cox transformation was performed [67]. Box-Cox transformed values were used later in the analysis of variance (ANOVA). This analysis represents an important tool in assessing the differences in selected elements according to the sampling campaign. With the assist of bivariate statistics, the correlation of content of chemical elements between individual moss samples according to the sampling campaign was initially established, which was then substantiated by statistical tests (*t*-test and *F*-test). The use of *t*-test and *F*-test allows the assessment of the statistical significance of differences between the accumulation values measured in different exposure periods. The enrichment ratio (ER) was used to assess the selected elements measured in various sampling campaigns. The ER is defined as the ratio of a grade of an element in some areas according to the natural abundance of the metal [68].

To investigate the degree of association of the chemical elements in the moss samples, the Pearson correlation coefficient I was used [69]. Using the correlation coefficient, the correlation degree (linear dependence) between two random variables or sets of random variables was calculated. A good association between the elements was assumed for r between 0.5 and 0.7, while a strong association was determined if r values ranged between 0.7 and 1.0. The elements with small values of the correlation coefficient were omitted from the matrices.

Based on the matrix of correlation coefficients, multivariate cluster, and R-mode factor analysis, the associations of chemical elements were revealed [70–72]. In this case, the values in a range of 0.5 to 0.7 were a good association, while the values in the range of 0.7 to 1.0 were a strong association. The multivariate statistical cluster and factor analyses were performed on 27 selected elements (Al, As, Co, Cs, Fe, Hf, Na, Rb, Sc, Ta, Th, Ti, U, V, Zr, Cd, Pb, Sb, Zn, Cr, Ni, Ba, K, Sr, Br, I, and rare-earth elements (RE)). Some elements were excluded from the analysis because they did not show a sensible connection with other chemical elements, or did not satisfy the criteria of dimension variables based on several observations. For factor analysis purposes, the variables were standardized to a zero mean [70,71,73]. Subsequently, based on characteristics of the 27 individual elements, five synthetic variables were designed (Factor 1—F1 to Factor 5—F5), which account for 81.8% of the total variability of the treated elements. The variables with factor loadings higher than 0.6 were assumed to contribute significantly to a given factor.

For the construction of maps showing the spatial distribution of factor scores and maps displaying the distribution of heavy metals in moss samples, universal kriging with the linear variogram interpolation method was applied [70]. For interpolation, the basic grid cell size was 1×1 km. Seven classes (0–10, 10–25, 25–40, 40–60, 60–75, 75–90, and 90–100) of the percentile values of the distribution of interpolation values were defined. The visualization (mapping) of data was performed using several software packages: Statistica 13 (StatSoft, Inc., Tuls, OK, USA), QGIS (#), and Surfer 17 (Golden Software, Inc., Golden, CO, USA).

3. Results and Discussion

The results of descriptive statistics to all assessed elements (aluminum, arsenic, barium, bromine, cadmium, cerium, cobalt, chromium, cesium, europium, iron, hafnium, iodine, potassium, lanthanum, sodium, nickel, lead, rubidium, antimony, scandium, samarium, strontium, tantalum, terbium, thorium, titanium, uranium, vanadium, zinc, and zirconium) in the 72 samples are shown in Table 1. The mean, median, minimum, and maximum contents were calculated for all 31 chemical elements (Al, As, Ba, Br, Cd, Ce, Co, Cr, Cs, Eu, Fe, Hf, I, K, La, Na, Ni, Pb, Rb, Sb, Sc, Sm, Sr, Ta, Tb, Th, Ti, U, V, Zn, and Zr) for the three sets of samples (Table 1). The results of the ANOVA and ER with respect to the sampling campaign are given in Table 2.

Table 1. Descriptive statistics of measurement according to sampling campaign; $n = 72$.

		2002			2005			2010		
	Unit	X_{BC}	Md	Min–Max	X_{BC}	Md	Min–Max	X_{BC}	Md	Min–Max
Al	%	0.40	0.38	0.083–1.8	0.37	0.36	0.15–2.6	0.24	0.24	0.11–0.68
As	mg/kg	0.86	0.80	0.12–8.0	0.69	0.68	0.18–4.3	0.51	0.48	0.23–1.9
Ba	mg/kg	52	54	14–260	52	53	18–190	50	50	14–360
Br	mg/kg	2.4	2.2	0.061–7.7	1.9	1.9	0.90–7.0	4.7	4.4	2.0–16
Cd	mg/kg	0.18	0.16	0.016–3.0	0.27	0.29	0.015–3.0	0.22	0.22	0.068–2.2
Ce	mg/kg	5.6	5.5	0.83–43	4.3	4.5	1.5–17	2.5	2.6	0.66–21
Co	mg/kg	1.2	1.1	0.24–14	1.2	1.1	0.42–5.3	0.77	0.83	0.27–2.9
Cr	mg/kg	7.8	7.9	2.3–120	7.0	6.8	2.1–82	6.7	6.5	2.5–35
Cs	mg/kg	0.40	0.38	0.097–1.7	0.34	0.32	0.13–2.3	0.20	0.20	0.093–0.90
Eu	µg/kg	110	110	27–480	71	66	6.3–480	71	79	8.9–260
Fe	%	0.25	0.24	0.042–1.7	0.23	0.22	0.10–0.81	0.18	0.19	0.089–0.54
Hf	mg/kg	0.29	0.26	0.050–3.8	0.22	0.21	0.076–1.1	0.17	0.17	0.085–0.70
I	mg/kg	1.2	1.2	0.36–2.8	1.6	1.7	0.64–3.7	1.6	1.6	0.51–2.9
K	%	0.83	0.84	0.29–1.8	0.75	0.75	0.47–1.4	0.66	0.66	0.36–1.0
La	mg/kg	2.5	2.3	0.50–22	2.4	2.3	0.97–9.1	1.3	1.4	0.62–9.0
Na	mg/kg	440	440	120–8700	360	360	140–1900	190	190	89–1000
Ni	mg/kg	2.6	2.5	0.089–24	6.1	5.8	1.8–43	4.0	4.3	1.0–55
Pb	mg/kg	5.8	6.0	1.5–37	7.4	7.6	0.10–47	4.5	4.6	1.9–22
Rb	mg/kg	11	11	5.0–47	9.4	9.8	4.0–29	6.6	6.6	2.2–21
Sb	mg/kg	0.19	0.20	0.040–1.4	0.15	0.15	0.044–0.92	0.09	0.09	0.044–0.22
Sc	mg/kg	0.85	0.74	0.12–6.8	0.72	0.67	0.30–4.6	0.46	0.44	0.16–1.9
Sm	mg/kg	0.44	0.45	0.072–3.4	0.37	0.35	0.14–2.0	0.27	0.27	0.11–1.5
Sr	mg/kg	31	32	12–140	32	34	13–140	34	34	12–120
Ta	µg/kg	91	93	13–790	78	77	31–330	38	40	18–170
Tb	µg/kg	60	62	9.4–560	55	53	15–250	34	36	12–190
Th	mg/kg	0.68	0.68	0.12–7.6	0.59	0.58	0.26–3.3	0.36	0.36	0.17–2.1
Ti	mg/kg	180	160	12–1400	230	220	86–1200	160	150	33–470
U	mg/kg	0.22	0.21	0.033–1.5	0.21	0.21	0.080–1.1	0.11	0.11	0.058–0.61
V	mg/kg	7.4	6.9	1.8–43	6.5	6.4	2.5–32	3.9	3.8	1.5–14
Zn	mg/kg	40	39	14–200	38	36	16–91	29	29	29
Zr	mg/kg	17	16	2.1–140	13	12	0.38–69	6.1	5.6	1.5–25

n—number of samples; X_{BC}—mean (Box-Cox transformed values used); Md—median; Min—minimum; Max—maximum.

Table 2. Statistical tests of differences and enrichment ratios according to sampling campaign (2002/2005/2010). Box-Cox transformed values were used for the analysis ($n = 72$).

		X_{BC}	X_{BC}	X_{BC}	t-test		t-test		t-test		F-test		ER	ER
	Unit	2002	2005	2010	2002/2005		2002/2010		2005/2010		2002/2005/2010		2005/2002	2010/2002
Al	%	0.40	0.37	0.24	−0.09	NS	4.59	*	5.23	*	15.0	*	0.92	0.61
As	mg/kg	0.86	0.69	0.51	2.19	*	5.25	*	2.98	*	13.3	*	0.79	0.59
Ba	mg/kg	52	52	50	−0.01	NS	0.43	NS	0.47	NS	0.1	NS	1.00	0.96
Br	mg/kg	2.4	1.9	4.7	1.96	NS	−8.68	*	−13.06	*	75.7	*	0.77	1.96
Cd	mg/kg	0.18	0.27	0.22	−1.33	NS	−1.34	NS	0.39	NS	1.2	NS	1.51	1.20
Ce	mg/kg	5.6	4.3	2.5	2.05	*	7.12	*	5.66	*	30.0	*	0.77	0.44
Co	mg/kg	1.2	1.2	0.77	−0.43	NS	3.93	*	5.04	*	12.9	*	1.04	0.67
Cr	mg/kg	7.8	7.0	6.7	0.77	NS	1.39	NS	0.59	NS	1.0	NS	0.90	0.86
Cs	mg/kg	0.40	0.34	0.20	0.94	NS	6.52	*	5.79	*	25.8	*	0.85	0.50
Eu	µg/kg	110	71	71	3.85	*	5.46	*	0.66	NS	13.4	*	0.62	0.62
Fe	%	0.25	0.23	0.18	0.18	NS	2.70	*	3.06	*	5.1	*	0.94	0.74
Hf	mg/kg	0.29	0.22	0.17	2.10	*	4.27	*	2.46	*	9.8	*	0.75	0.60
I	mg/kg	1.2	1.6	1.6	−4.55	*	−4.36	*	0.71	NS	14.2	*	1.33	1.35
K	%	0.83	0.75	0.66	1.30	NS	4.16	*	4.47	*	11.4	*	0.90	0.79
La	mg/kg	2.5	2.4	1.3	0.29	NS	5.65	*	6.21	*	23.4	*	0.95	0.53
Na	mg/kg	440	360	190	1.35	NS	7.56	*	6.98	*	36.0	*	0.83	0.43
Ni	mg/kg	2.6	6.1	4.0	−6.85	*	−3.92	*	3.15	*	24.8	*	2.34	1.54
Pb	mg/kg	5.8	7.4	4.5	−0.76	NS	2.53	*	2.71	*	4.3	*	1.26	0.77
Rb	mg/kg	11	9.4	6.6	3.21	*	7.86	*	4.99	*	33.3	*	0.82	0.58
Sb	mg/kg	0.19	0.15	0.09	1.87	NS	7.57	*	6.41	*	31.5	*	0.80	0.48
Sc	mg/kg	0.85	0.72	0.46	0.64	NS	5.31	*	5.38	*	18.5	*	0.85	0.54
Sm	mg/kg	0.44	0.37	0.27	1.17	NS	3.78	*	2.99	*	8.3	*	0.84	0.62
Sr	mg/kg	31	32	34	−0.40	NS	−0.55	NS	−0.20	NS	0.2	NS	1.05	1.10
Ta	µg/kg	91	78	38	1.02	NS	8.19	*	8.78	*	46.3	*	0.86	0.42
Tb	µg/kg	60	55	34	0.51	NS	4.46	*	4.56	*	13.6	*	0.92	0.57
Th	mg/kg	0.68	0.59	0.36	0.79	NS	5.56	*	5.52	*	20.6	*	0.88	0.52
Ti	mg/kg	180	230	160	−2.59	*	0.91	NS	4.52	*	7.7	*	1.28	0.93
U	mg/kg	0.22	0.21	0.11	0.25	NS	5.87	*	6.59	*	25.6	*	0.93	0.50
V	mg/kg	7.4	6.5	3.9	0.71	NS	6.59	*	6.37	*	27.1	*	0.88	0.53
Zn	mg/kg	40	38	29	0.84	NS	4.64	*	4.26	*	13.4	*	0.94	0.72
Zr	mg/kg	17	13	6.1	2.65	*	8.38	*	5.37	*	32.7	*	0.78	0.37

X_{BC}—mean (Box-Cox transformed values); ER—enrichment ratio; NS—nonsignificant; *—significant.

In Table 1, the mean values, median values, and the interval ranges for the contents of all 31 selected elements used for comparison between moss surveys measurements are introduced. Generally, the medians of the elements decreased according to the sampling campaign. However, there are some exceptions. The median content of Ni content from 2005 (5.8 mg/kg) and 2010 (4.3 mg/kg) was higher than in 2002 (2.5 mg/kg), which is related to reactivation and increasing the production capacity of the ferronickel smelter near Kavadarci in 2004 [33].

A similar trend can be observed for Cd and Pb. The higher values of these two elements in 2005 than in 2002 can be explained by pollution associated with the polluted soil and the existence of an open slag waste dump in the vicinity of the abandoned lead-zinc smelter near the town of Veles [24]. In contrast, the lower median values of Cd and Pb measured in 2010 compared with those obtained in 2002 and 2005 can be explained by the closure of the lead-zinc smelter in Vales in 2003.

The presence of iodine and bromine as an association of these two elements (halogens) is primarily due to marine influence. Along the valley, the Vardar River, the Jugo wind blows from the northern parts of Africa across the Aegean Sea and brings with itself particles enriched with these two elements, which could be subsequently trapped in mosses [74]. The titanium also does not follow a declining trend over the years. The values of Ti from 2005 are higher than in 2002, which can be related to the increased mining activities in the copper flotation plant near Radoviš [22].

Al, As, Ba, Ce, Co, Cs, Eu, Fe, Hf, La, Na, Rb, Sc, Sm, Ta, Tb, Th, U, V, and Zr represent the typical crustal composition. However, it cannot be overlooked the significant fall of some median values of elements according to the sampling campaign. One possible explanation in this case study could be related to the intensity of precipitation and leaching of trapped particles from the mosses, reducing the values by as much as 20%, according to previous studies [75]. In eight regions in Macedonia, the precipitation increased between 2005 and 2010, which further confirms the aforementioned study [76,77].

The differences in the measurements of individual elements, according to the sampling campaign, were supported by the ANOVA analysis. The *t*-test was used to compare the values of the elements obtained in two different sampling campaigns (Table 2). There were no significant differences between the contents of particular elements measured in 2002 and 2005, which means that all mining and metallurgy activities continuously operated within the same capacity during that period; there were also no natural changes. There were significant differences between the contents of elements measured in 2002 vs. 2010 and the contents of elements obtained in 2005 vs. 2010, respectively, related to either natural or anthropogenic influences. This was then also confirmed by the *F*-test, which showed that there are significant differences in element values in all three moss surveys.

The enrichment ratio (ER) for all selected 31 elements is given in Table 2. The highest enrichment ratio is found for Ni (2.34) in moss samples from 2002 compared with those from 2005 (Table 2). High values of Ni were a consequence of the increased workload in ferronickel smelter in the vicinity of Kavadarci in 2004. The highest enrichment ratio was found for Br (1.96), which is a marine halogen. The Aegean Sea is less than 60 km from the southern state border, and the wind blowing from the northern parts of Africa across the Aegean Sea brings halogen-enriched particles. Over the years, the wind may be strengthened, increasing the content of Br in mosses [78].

For a better visibility, the matrix of correlations coefficients was divided into two matrices. Table A1 is a matrix of correlation coefficient of 16 selected naturally distributed elements; Table A2 is a matrix of correlation coefficients of 11 selected elements. In both matrices, the Box-Cox transformed values were used.

Three geogenic, one anthropogenic association, and one mixed (geogenic-anthropogenic) geochemical associations were established as a result of a visual inspection of the similarities of the spatial distribution of element patterns, a comparison of basic statistical parameters, the correlation coefficient matrices (Tables A1 and A2), and the results of multivariate analyses (cluster and factor analysis). Variables or samples with similar behavior were combined into the cluster according to the Pearson correlation coefficient. The results are presented in a dendrogram, which was performed to show the results of the hierarchical cluster analysis of 27 selected elements (Figure 4); and with factor analysis, where the grouping of the elements into five groups was performed (Table 3). The same classification trend into the factor score or cluster can be seen in both analyses. Four elements were excluded from the clustering because of the absence of similarity linkage.

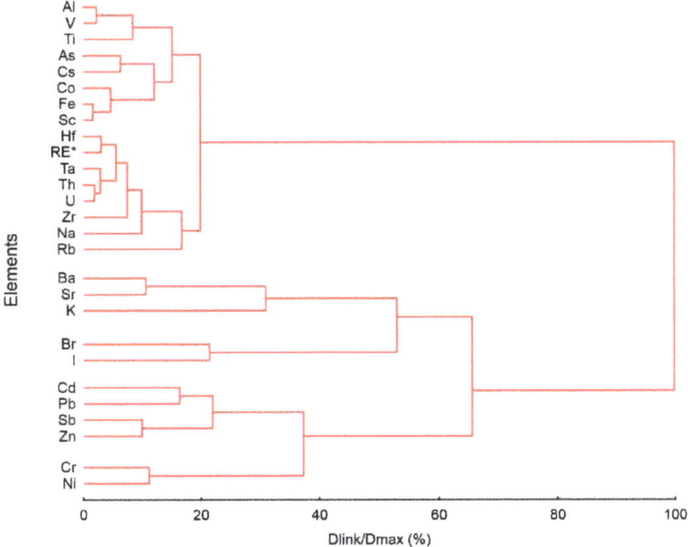

Figure 4. Cluster analysis dendrogram—element relationship (*n* = 216, 27 selected elements). RE*—Average of the standardized values of analyzed rare-earth elements: Ce, Eu, La, Sm, and Tb.

The obtained factors were also compared according to the sampling campaigns (2002/2005/2010) (Figure 5). The largest changes were observed in Factor 1 and Factor 5. Factor 1 falls sharply according to the sampling campaign; the opposite trend is observed at Factor 5. Both factors illustrate the geogenic association of chemical elements, which is related to the climatic variables. Factor 3 follows a concave pattern with the highest values obtained in 2005, which coincides with the reactivation of the ferronickel smelter in Kavadarci in 2004. Factor 2 slightly declined, which can be related to the closure of Pb-Zn mine in Veles in 2003. The geogenic association Factor 4 shows almost no changes, which means there were no significant natural changes.

The matrix of factor loadings is shown in Table 3. Combined, the five factors explain 81.8% of the variability of the threatening elements. Factor 1 (F1) represents the strongest factor, with 47.3% of the total variability. This group of elements represents the geogenic group of elements. Factor 2 (F2) is the second strongest factor, with 10.4% of the total variability. This factor is associated with elements Pb, Zn, and Cd and represents the anthropogenic elements. The third factor (F3) explains 8.6% of total variability and includes elements of Cr and Ni, whose distribution can be natural or anthropogenic. Factor (F4) represents 8.5% of total variability and includes elements that show natural distribution. Factor 5 (F5) explains 7% of total variability and includes Br and I.

Table 3. Matrix of dominant rotated factor loadings normal values (n = 216, 27 selected elements).

	F1	F2	F3	F4	F5	Comm
Al	**0.90**	0.08	0.19	0.14	0.17	91.0
As	**0.78**	0.32	0.21	−0.01	0.07	76.7
Co	**0.73**	0.20	0.56	0.15	0.04	91.1
Cs	**0.87**	0.18	0.14	0.14	−0.01	83.5
Fe	**0.84**	0.15	0.36	0.20	0.19	93.2
Hf	**0.90**	0.04	0.13	0.24	0.15	90.1
Na	**0.80**	0.11	0.11	0.33	−0.18	80.3
Rb	**0.69**	0.18	−0.16	0.40	−0.13	71.3
Sc	**0.90**	0.11	0.29	0.12	0.06	93.1
Ta	**0.94**	0.10	0.07	0.16	−0.03	92.1
Th	**0.91**	0.12	0.09	0.24	0.08	92.0
Ti	**0.72**	0.00	0.27	0.21	0.27	70.6
U	**0.89**	0.15	0.11	0.27	0.04	90.5
V	**0.87**	0.19	0.23	0.03	0.07	85.5
Zr	**0.89**	0.06	0.03	0.17	−0.06	82.6
RE*	**0.90**	0.08	0.17	0.24	0.16	91.7
Cd	−0.08	**0.82**	0.13	0.12	0.11	73.2
Pb	0.22	**0.83**	0.02	−0.06	0.18	77.8
Sb	0.59	**0.64**	0.19	−0.10	−0.11	82.9
Zn	0.33	**0.73**	0.33	0.12	−0.20	81.3
Cr	0.53	0.22	**0.69**	0.05	0.12	81.8
Ni	0.18	0.25	**0.86**	0.05	0.21	87.9
Ba	0.43	−0.00	−0.03	**0.75**	0.26	81.9
K	0.20	0.11	0.01	**0.68**	−0.39	67.0
Sr	0.31	−0.02	0.22	**0.70**	0.24	69.2
Br	−0.04	0.03	0.05	0.08	**0.81**	66.8
I	0.24	0.11	0.19	−0.00	**0.72**	63.1
Prp.Totl	47.3	10.4	8.6	8.5	7.0	81.8
Expl.Var	12.78	2.81	2.33	2.29	1.88	
EigenVal	15.34	2.50	0.98	1.34	1.92	

F1–F5—factor loadings; Com—communality in %; Prp.Totl—principal total variance in %; Expl.Var—explained variance; EigenVal—eigenvalues; Dominant values are bolded; RE*—an average of standardized values of analyzed rare-earth elements: Ce, Eu, La, Sm, and Tb.

Association of Al, As, Co, Cs, Fe, Hf, Na, Rb, Sc, Ta, Th, Ti, U, V, Zr, and RE (Factor 1) represents the typical crustal components that are significantly influenced by the mineral particles that are carried into the atmosphere and later trapped in the moss samples by the wind. Generally, the highest content of these elements was detected in samples collected in the vicinity of Precambrian and Paleozoic shales. Shales decomposed under the influence of various weather conditions, and the minerals of these elements are released into the environment. The annual precipitation is the lowest in the regions of Tikveš, Štip, and Ovče Pole, which leads to erosion of decomposed shales by wind and significantly affects the spread of mineral particles [6]. In other words, the distribution of elements from these groups is independent of urban and industrial areas, but depended on the geological background. The median values of some of these elements have changed over the sampling campaign (Table 1), mostly decreasing, which can also be seen from the spatial distribution of Factor 1 (Figures 5 and 6).

This can be supported by the presence of Al in this group. Aluminum compounds are insoluble in water, and the total content of Al in various biological systems originates from dust pollution [79]. Similarly, the presence of iron in the samples originated from dust particles. The highest content of Fe was found in samples collected from the central part of the country, in an area with the smallest annual precipitation and low vegetation, where the land is susceptible to various erosion factors. For the same reasons, the influence of the steel smelter in Skopje on the content of these elements in the samples was not to be observed.

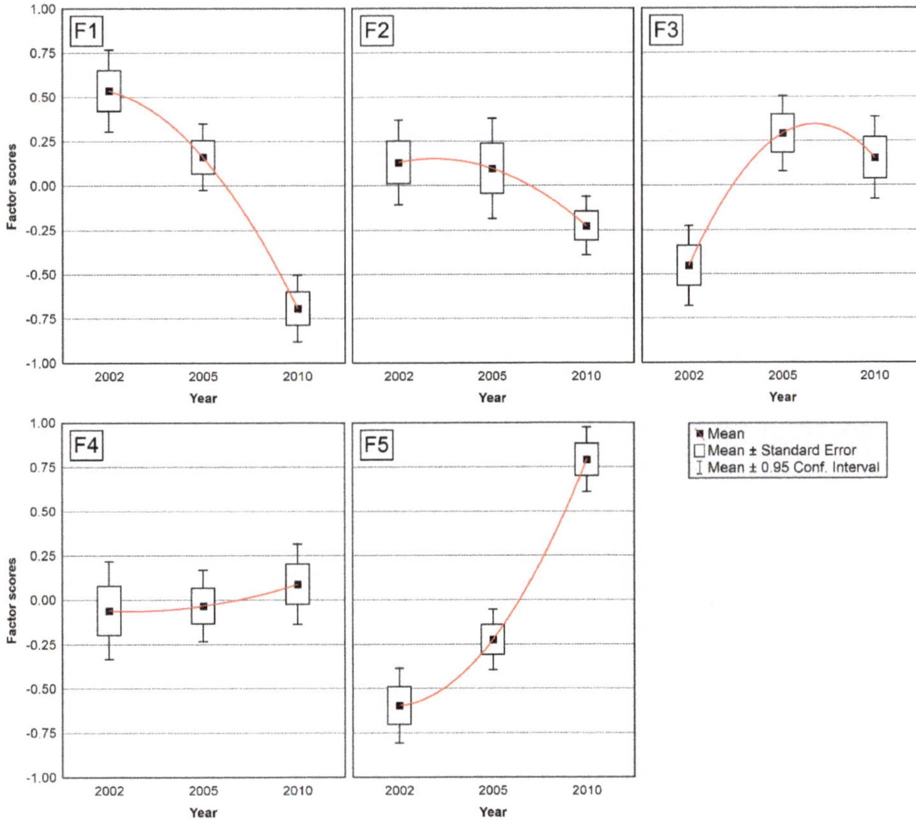

Figure 5. Box plots of factor scores (st. values) according to the sampling campaign. F1–F5—factor loadings.

There was a declining trend in the content of elements grouped into Factor 1, according to the sampling campaign (Figure 5). In 2002, the highest concentrations of the Factor 1 elements were found in moss samples, collected in the regions of Bitola, Strumica, Radoviš, and Skopje. The high values of these elements (especially uranium) in the samples collected in the Serbo-Macedonian massif and the Pelagonia massif (Figure 6) can be explained by the existence of uranium ore deposits in these regions [44,59]. The second peak of these elements (mostly related to U and Th) was detected in the samples taken in the southwestern part of the country near Bitola, which can be related to fly ash emissions from a thermoelectric power plant in Bitola, where lignite was used as fuel. It is also related to transboundary pollution from several thermoelectric power plants in northern Greece [2]. The uranium-rich granite deposits in the Strumica region can also affect the presence of high values of uranium [44,59]. During the exploitation of these granites, uranium-reach dust particles may be released into the environment in considerable amounts, trapped by the moss [2].

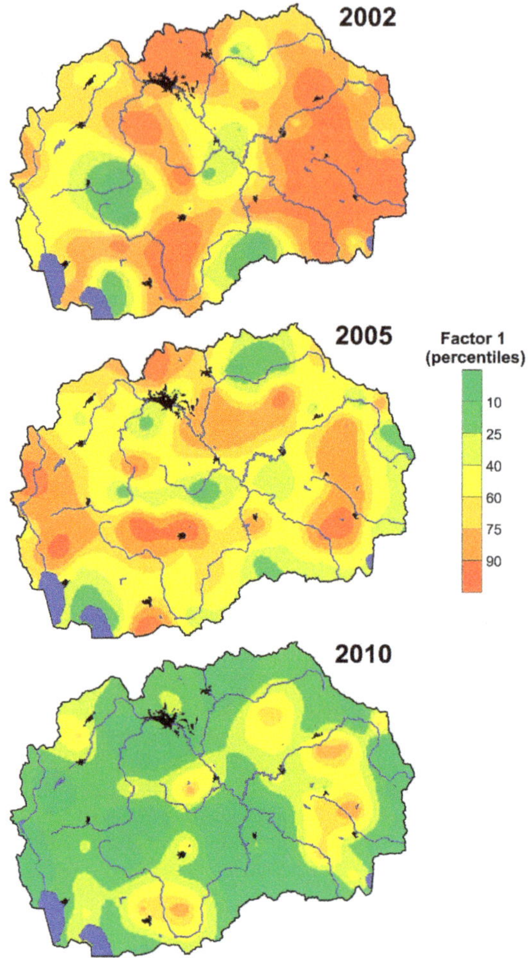

Figure 6. Spatial distribution of F1 factor scores (Al-As-Co-Cs-Fe-Hf-Na-Rb-Sc-Ta-Th-Ti-U-V-Zr-RE*).

The high representation of these elements, especially Al, Sc, Ti, V in the vicinity of Radoviš, can be related to the activities of the copper mine and flotation plant Bučim [27,28,37,80]. High values of these elements (Sb, Fe, and Co) near Tetovo and Kavadraci are related to the previously active ferrochrome smelter in Tetovo and current activity near Kavadarci (ferronickel smelter) [2,79,81].

In 2005, the anomalies were no longer so clear. The anomaly can be seen along the Radika River in Western Macedonia where Neogene clastic sediments predominate [44]. Higher contents of elements from that group were also found in the vicinity of Galičica, due to the Mesozoic carbonate rocks that occur in these areas. Furthermore, areas with high contents of the element group in Factor 1 were found in the area south of Skopje, and in the vicinity of Prilep both areas with Proterozoic metamorphic rocks from the Pelagonian massif [8]. The median values are mostly lower than the median values of the same elements measured in 2002.

The median values for these elements were lowest in 2010 (Figure 5). According to the State statistical office in 2011 and 2013, the precipitation in eight regions in Macedonia was greater in 2010 than in 2005. However, the central part of the country is characterized by low plant cover and the lowest amount of precipitation, which means these areas are most exposed to erosion, and suspension of soil material [22].

The spatial distribution of Factor 2 (Cd-Pb-Sb-Zn) was not influenced by lithological background and is mainly connected with a lead-zinc smelter (Figure 7). Factor 2 was directly related to the dust from the flotation tailings at the mines and from the soils that are spread into the atmosphere by the winds, leading to the atmospheric distribution of these elements [3,39,41–43]. Air pollution decreased over the years, even though the drop in the content of elements in Factor 2 was not as clear as in Factor 1 (Figure 5). In 2002, the enrichment with these elements can be seen along the Vardar River valley, which extends from Skopje to Veles and continues southeastward toward Greece. Due to the frequent flow of air masses in both directions in summer and winter, there was an accelerated diversification of emissions from industry. In 2002, the pollution was mostly connected with the lead and zinc smelter plant in Veles–which had operated until 2003 and with the steel-work smelter in Skopje [2].

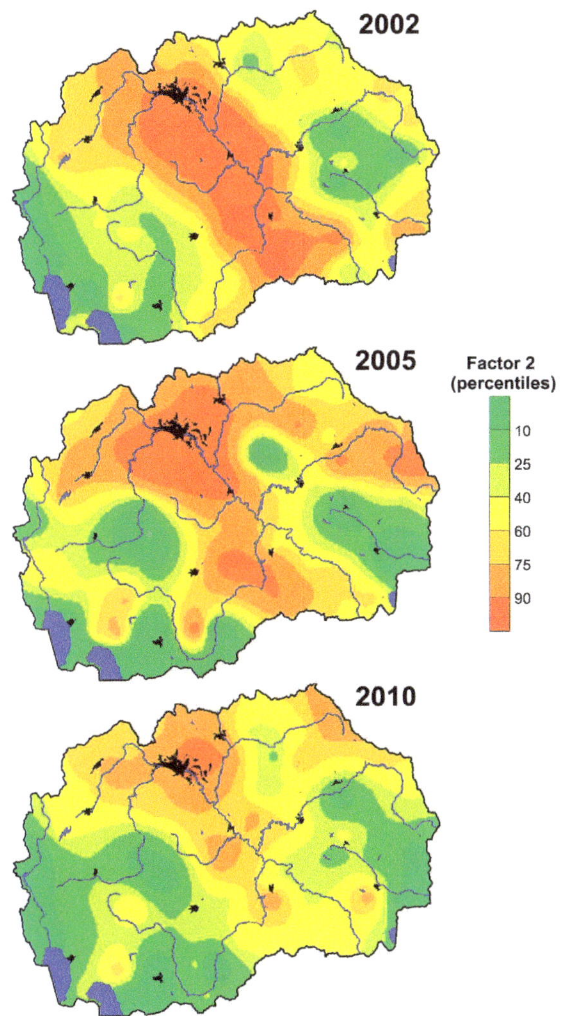

Figure 7. Spatial distribution of F2 factor scores (Cd-Pb-Sb-Zn).

In 2005, the ER of the elements from Factor 2 is still connected with pollution by the lead and zinc smelter plant in Veles despite its closure in 2003. The high values in 2005 for these elements are also partly associated with the steel-work smelter in Skopje [32]. Leaded gasoline, which was widely used during that period, could also be a possible source of lead in the surface-atmosphere of Macedonia [32]. In contrast with the moss survey in 2002, enrichment with these elements in 2005 was also observed in the area of Bitola in the eastern part of the country. According to the previous study, the Neogene vulcanites were characterized by the natural enrichment of Cd in the region of Berovo-Vladimirovo [42].

In 2010, due to the reactivation of Pb-Zn mines (Sasa, Toranica, Zletovo), the enrichment of Cd and Zn were observed in the eastern part of Macedonia. In the period of 2001/2002 until 2006/2007, the mines were not active, but they were reactivated in 2006 (Sasa) and 2007 (Toranica, Zletovo). In the area of the mines, there were millions of tons of waste material, which can be dispersed into the biosphere by the wind [41–43,82]. Furthermore, the enrichment in 2010 was also observed in the area of the steel-work smelter in Skopje. In the surveys from 2010, the contents of Cd and Pb decreased because of the reduced use of leaded gasoline for cars [8].

Factor 3 (Ni, Cr) contains elements that are usually associated with air pollution, but also influenced by a natural factor, such as the lithological background (Figure 8). The enrichment of nickel is geogenic, reaching the locality of Groot near the town of Vales in the central part of the country. This location also has Paleogene flysch sediments and Neogene sediments, whose particles can be spread by the wind along the valleys of the Vardar River and Crna River [53]. Similar findings have also been made in previous studies in Macedonia [25,30,81]. The main anthropogenic source of these elements is related to the Kavadarci region and includes the ferronickel smelter, the slag dump, and the open pit nickel mine [25,30,31,82]. Generally, the high chromium content belongs to the region near Tetovo, an area of a previous ferrochromium smelter and a slag deposit near the high-melting ferrochromium plant [2].

There were fluctuations in the contents of Ni and Cr, according to the sampling campaign (Figure 5). The lowest contents were observed in 2002; the highest was in 2005. In samples from moss surveys in 2002, the element contents from the association Factor 3 were related to the geogenic distribution associated with Paleogene flysch sediments and Neogene sediments. The high content was also be observed in the area of Tetovo and Kavadarci, which were associated with a previous ferrochrome smelter and the abandoned ferronickel smelter, respectively. Values for Ni in 2005 were much higher than the values for 2002, which was related to reactivation and increasing production capacity of the ferronickel smelter near Kavadarci in 2004 [30,33,80]. Since 2005, the factory has increased its ore processing capacity, as it also extracts material from Albania, Turkey, and Indonesia [34]. However, the contents of Ni and Cr in the mosses collected in 2010 slightly decreased, which can be linked to a renewed decline in the production capacity at the smelter.

Factor 4 (Ba, K, Sr) is the association of elements that are naturally distributed and whose distribution is not related to industrial and urban activities (Figure 9). The contents of these elements are practically unaltered. The elements are enriched in areas of felsic volcanic rocks (andesite) and their pyroclastic rocks. The contents of these elements slightly increased according to the sampling campaign. In 2002 the enrichment was observed in Kratovo, Ovče Pole, and Kavadarci, where Neogene (Ng) magmatic rock and Paleogene cluster dominates. Another maximum was linked to the southern Macedonia (Kožuf and Mariovo regions) and Strumica area where rocks of Quaternary (Q) sediments dominated. In 2005, the highest values of these elements were found in the area of Kratovo (rocks of Neogene volcanism) [44]; in 2010, the spatial distribution was similar to the distribution from 2002.

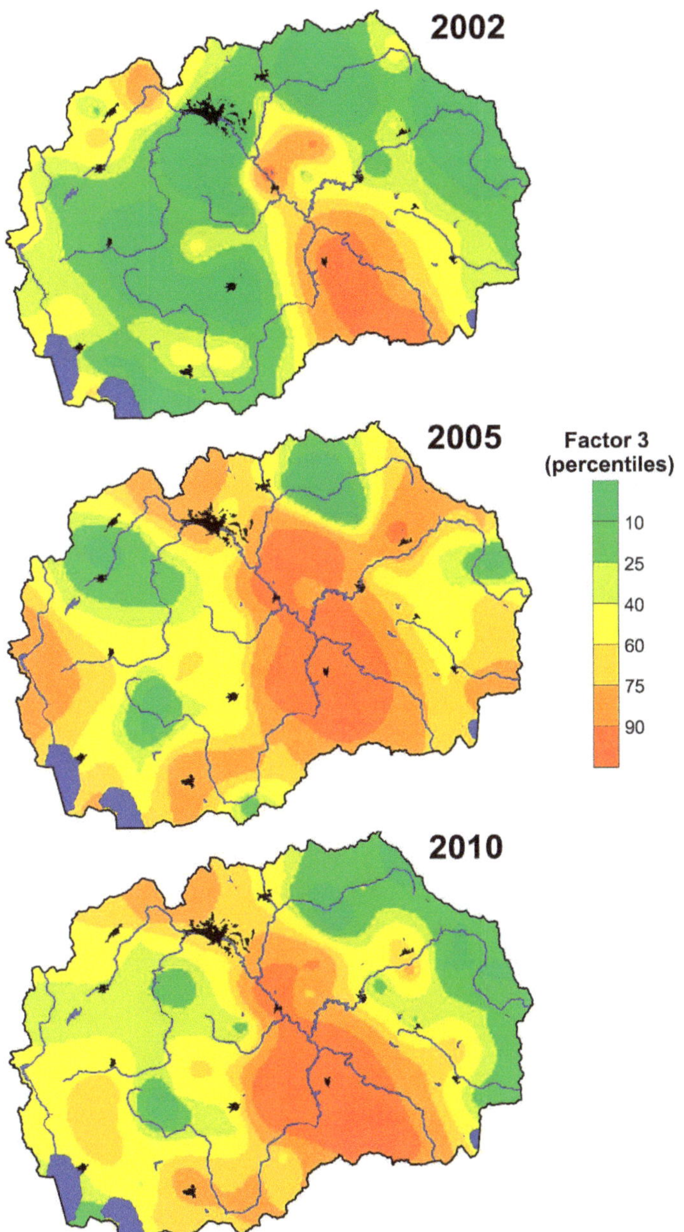

Figure 8. Spatial distribution of F3 factor scores (Ni-Cr).

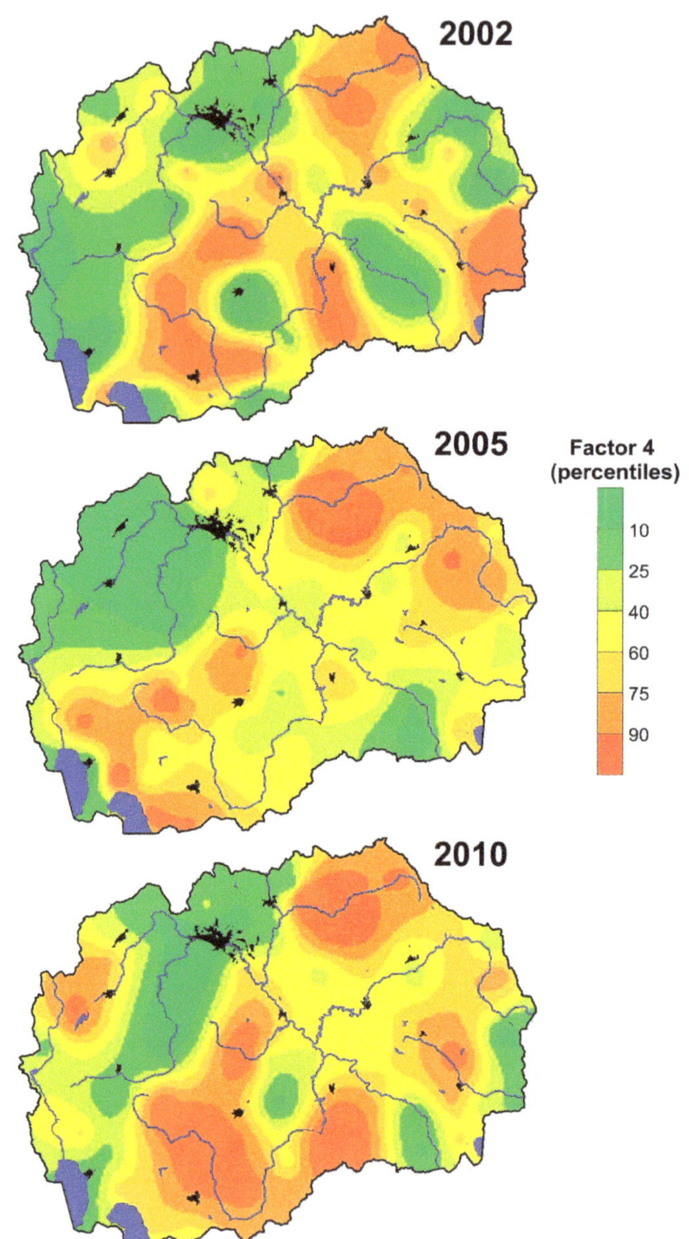

Figure 9. Spatial distribution of F4 factor scores (Ba-K-Sr).

Factor 5 (Br, I) includes elements primarily affected by marine influence [74]. Along the Vardar River valley during the winter period, the Jugo wind blows from the northern parts of Africa across the Mediterranean and the Aegean Sea. Consequently, an exponential decrease of Br and I with the distance from the coastline was observed. This phenomenon has already been studied in Norway, where high contents of halogens were observed in soils and mosses corresponding to an exponential decline with distance from the coastline [74,83]. The limestone and dolomite (Paleozoic and Mesozoic carbonates) present in the western part of Macedonia are enriched in iodine, which also explains the elevated values of these elements in the mosses [32]. In 2010, the contents of Br and I were the highest, which means that the wind was stronger in 2010 (Figure 10).

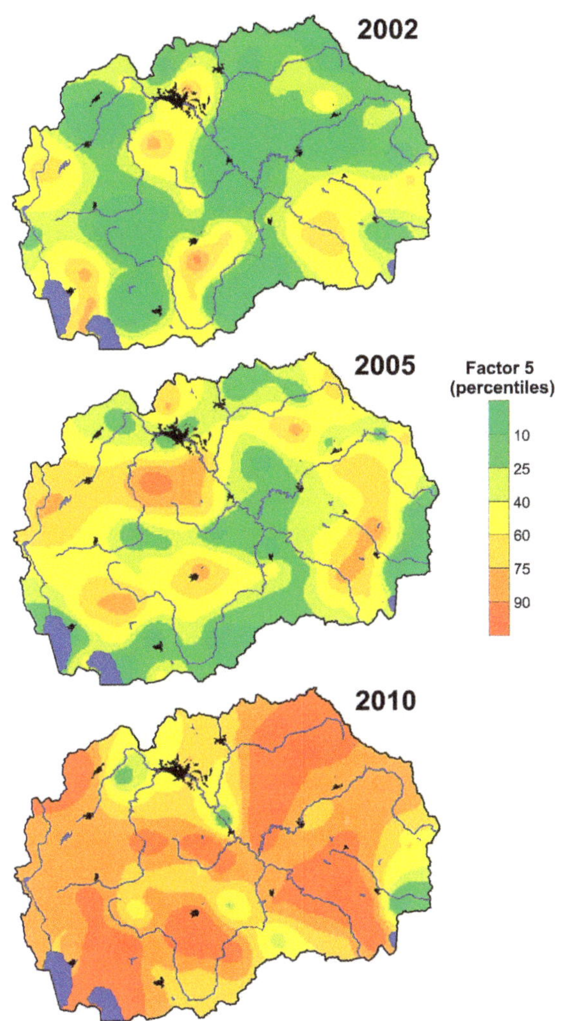

Figure 10. Spatial distribution of F5 factor scores (Br-I)—normal values.

4. Conclusions

The moss analysis is a valuable method for monitoring atmospheric deposition of trace elements in Macedonia mainly because it provides a cheap, effective alternative to deposition analysis for identification of areas at risk from high atmospheric deposition fluxes of heavy metals and can play an important role in identifying spatial and temporal trends in atmospheric heavy metal pollution across the country. In this contribution, the results of three air pollution moss surveys of heavy metals are presented and compared. The first systematic study of heavy metals atmospheric pollution using the moss technique was concluded in 2002 [1,2]. This study was followed by two more samplings in 2005 [32] and 2010 [6,22], collecting moss samples from the same locations. In this study, the results of chemical analyses obtained from the aforementioned studies were statistically processed using bivariate and multivariate statistical techniques. By the comparison of all three moss surveys, it can be concluded that some of the potentially toxic elements (Cd, Co, Ni, Pb, Ti, Zn) have increased in content in moss samples from 2002 to 2005, but decreased in 2010. Increased values were also observed for Br in 2005 compared to 2010, and for I in 2005 and 2010 compared to 2002.

The largest anthropogenic impact of air pollution with heavy metals was found in the vicinity of the ferronickel smelter near Kavadarci (Ni and Cr), the abandoned lead-zinc smelter near Vales, and the lead and zinc mines in the vicinity of Macedonska Kamenica, Probištip, and Kriva Palanka (Cd, Pb, and Zn). According to factor analysis, three geogenic associations following the lithology and the composition of the soil (Factor 1, which includes Al, As, Co, Cs, Fe, Hf, Na, Rb, Sc, Ta, Th, Ti, U, V, Zr, and RE; Factor 4 with Ba, K, and Sr; Factor 5 with I and Br), one geogenic-anthropogenic association (Factor 3 with Cr and Ni), and one anthropogenic (Factor 2, which includes Cd, Pb, Sb, and Zn) association were obtained.

According to the sampling campaign (2002/2005/2010), the content of elements associated with Factor 1 and Factor 5 changed the most. Factor 1 decreased considerably according to the sampling campaign, while the opposite trend was observed for Factor 5. These two factors show geogenic distribution, mostly related to natural background and weather conditions. The contents of the elements combined in Factor 3 increased from moss surveys from 2002 to 2005 and fall slightly in 2010. This coincides with the reactivation of a ferronickel smelter in Kavadarci in 2004. Factor 2 slightly declined, according to the sampling campaign, which may be related to the closure of the Pb-Zn mine in Vales in 2003. The geogenic association in Factor 4 shows almost no changes.

The largest anthropogenic impact of air pollution with potentially toxic metals was the ferronickel smelter plant in the vicinity of Kavadarci (Ni and Cr), as well as three reactivated lead and zinc mines near the towns of Probištip, Makedonska Kamenica and Kriva Palanka (Cd, Pb, and Zn). The distribution of analyzed elements in the Republic of Macedonia was also related to the lithology and the composition of the soil.

This work is essential for monitoring future trends at a high spatial resolution and provides a useful tool for possible modeling the atmospheric deposition fluxes. Environmental monitoring programs, such as the moss survey are also appropriate to regulatory bodies of the country, with the aim to prevent the quality of the environment from deteriorating or ensure that its quality is improved.

Author Contributions: Conceptualization, T.S.; Data curation, R.Š. and K.B.A.; Investigation, L.B., T.S., M.F. and K.B.A.; Methodology, M.F.; Software, R.Š.; Writing—original draft, R.Š.; Writing—review & editing, L.B. and T.S. All authors have read and agreed to the published version of the manuscript.

Funding: This research received no external funding.

Conflicts of Interest: The authors declare no conflict of interest.

Appendix A

Table A1. Matrix of correlation coefficients (n = 216), a group of 16 selected principally natural distributed elements).

	Al	As	Co	Cs	Fe	Hf	Na	Rb	Sc	Ta	Th	Ti	U	V	Zr	RE*
Al	1.00															
As	**0.76**	1.00														
Co	**0.79**	**0.73**	1.00													
Cs	**0.83**	**0.83**	**0.77**	1.00												
Fe	**0.88**	**0.79**	**0.90**	**0.83**	1.00											
Hf	**0.87**	**0.74**	**0.78**	**0.82**	**0.90**	1.00										
Na	**0.77**	<u>0.64</u>	**0.70**	**0.74**	**0.76**	**0.78**	1.00									
Rb	<u>0.63</u>	<u>0.54</u>	<u>0.51</u>	**0.72**	<u>0.58</u>	0.64	<u>0.67</u>	1.00								
Sc	**0.89**	**0.80**	**0.89**	**0.86**	**0.96**	**0.90**	**0.79**	0.59	1.00							
Ta	**0.87**	**0.74**	**0.78**	**0.85**	**0.84**	**0.90**	**0.82**	**0.72**	**0.89**	1.00						
Th	**0.89**	**0.78**	**0.77**	**0.88**	**0.86**	**0.90**	**0.80**	**0.72**	**0.88**	**0.94**	1.00					
Ti	**0.84**	<u>0.57</u>	<u>0.69</u>	<u>0.66</u>	**0.77**	**0.73**	<u>0.62</u>	<u>0.51</u>	**0.75**	**0.70**	**0.71**	1.00				
U	**0.88**	**0.78**	**0.77**	**0.86**	**0.85**	**0.88**	**0.84**	**0.71**	**0.87**	**0.92**	**0.95**	**0.72**	1.00			
V	**0.94**	**0.76**	**0.80**	**0.80**	**0.85**	**0.79**	**0.74**	<u>0.58</u>	**0.89**	**0.82**	**0.81**	**0.79**	**0.81**	1.00		
Zr	**0.81**	<u>0.67</u>	**0.70**	**0.77**	**0.80**	**0.87**	**0.78**	<u>0.66</u>	**0.83**	**0.87**	**0.84**	<u>0.65</u>	**0.82**	**0.78**	1.00	
RE*	**0.91**	**0.77**	**0.81**	**0.83**	**0.91**	**0.92**	**0.77**	<u>0.68</u>	**0.91**	**0.88**	**0.92**	**0.75**	**0.90**	**0.84**	**0.83**	1.00

n = total number of samples from the three sampling campaigns; Values in the range of 0.5–0.7 (good association) are underlined, and in the range 0.7–1.0 (strong association) are bolded; Box-Cox transformed values used. RE*–an average of standardized values of analyzed rare-earth elements: Ce, Eu, La, Sm, and Tb.

Table A2. Matrix of correlation coefficient (n = 216), a group of 11 selected elements, natural and anthropogenic elements.

	Ba	Br	Cd	Cr	I	K	Ni	Pb	Sb	Sr	Zn
Ba	1.00										
Br	0.16	1.00									
Cd	0.04	0.11	1.00								
Cr	0.29	0.19	0.21	1.00							
I	0.25	<u>0.53</u>	0.12	0.34	1.00						
K	<u>0.50</u>	−0.15	0.06	0.15	−0.07	1.00					
Ni	0.16	0.18	0.33	**0.71**	0.41	0.01	1.00				
Pb	0.12	0.08	<u>0.57</u>	0.32	0.25	0.01	0.32	1.00			
Sb	0.20	−0.07	0.41	<u>0.50</u>	0.19	0.17	0.38	<u>0.61</u>	1.00		
Sr	**0.72**	0.14	0.10	0.34	0.19	0.44	0.29	0.12	0.17	1.00	
Zn	0.17	−0.06	<u>0.51</u>	0.49	0.12	0.33	0.45	<u>0.61</u>	**0.74**	0.15	1.00

Values in the range 0.5–0.7 (good association) are underlined, and in the range 0.7–1.0 (strong association) are bolded; Box-Cox transformed values used.

References

1. Barandovski, L.; Cekova, M.; Frontasyeva, M.V.; Pavlov, S.S.; Stafilov, T.; Steinnes, E.; Urumov, V. Air Pollution Studies in Macedonia Using the Moss Biomonitoring Technique, NAA, AAS, and GIS technology. *Frank Lab. Neutron Phys.* **2006**, preprint.
2. Barandovski, L.; Cekova, M.; Frontasyeva, M.V.; Pavlov, S.S.; Stafilov, T.; Steinnes, E.; Urumov, V. Atmospheric deposition of trace element pollutants in Macedonia studied by the moss biomonitoring technique. *Environ. Monit. Assess.* **2008**, *138*, 107–118. [CrossRef] [PubMed]
3. Angelovska, S.; Stafilov, T.; Šajn, R.; Balabanova, B. Geogenic and anthropogenic moss responsiveness on lithological elements distribution around Pb-Zn ore deposit. *Arch. Environ. Contam. Toxicol.* **2016**, *70*, 487–505. [PubMed]
4. Loppi, S.; Giomarelli, B.; Bargagli, R. Lichens and mosses as biomonitors of trace elements in a geothermal area (Mt. Amiata, Central Italy). *Cryptogr. Mycol.* **1999**, *20*, 119–126. [CrossRef]

5. Kłos, A.; Ziembik, Z.; Rajfur, M.; Dołhańczuk-Śródka, A.; Bochenek, Z.; Bjerke, J.W.; Tommervik, H.; Zagajewski, B.; Ziółkowski, D.; Jerz, D.; et al. Using moss and lichens in biomonitoring of heavy-metal contamination of forest areas in southern and north-eastern Poland. *Sci. Total Environ.* **2018**, *627*, 438–449. [CrossRef] [PubMed]

6. Barandovski, L. Use of passive biomonitoring and neutron activation analysis for deposition monitoring of trace elements in the air in the Republic of Macedonia. Ph.D. Thesis, Ss Cyril and Methodius University, Skopje, North Macedonia, 2014; 102p. (In Macedonian)

7. Lazo, P.; Stafilov, T.; Qarri, F.; Allajbeu, S.; Bekteshi, L.; Frontasyeva, M.; Harmens, H. Spatial distribution and temporal trend of airborne trace metal deposition in Albania studied by moss biomonitoring. *Ecol. Indic.* **2019**, *101*, 1007–1017. [CrossRef]

8. Stafilov, T.; Šajn, R.; Barandovski, L.; Bačeva, K.A.; Malinovska, S. Moss biomonitoring of atmospheric deposition study of minor and trace elements in Macedonia. *Air Qual. Atmos. Health* **2018**, *11*, 137–152. [CrossRef]

9. Szczepaniak, K.; Biziuk, M. Aspects of the biomonitoring studies using mosses and lichens as indicators of metal pollution. *Environ. Res.* **2003**, *93*, 221–230. [CrossRef]

10. Wróbel, Ł.; Dołhańczuk-Śródka, A.; Kłos, A.; Ziembik, Z. The activity concentration of post-Chernobyl ^{137}Cs in the area of the Opole Anomaly (southern Poland). *Environ. Monit. Assess.* **2015**, *187*, 1–7. [CrossRef]

11. Dołhańczuk-Śródka, A.; Ziembik, Z.; Kříž, J.; Hyšplerova, L.; Wacławek, M. Pb-210 isotope as a pollutant emission indicator. *Ecol. Chem. Eng. S* **2015**, *22*, 73–81.

12. Groet, S.S. Regional and local variations in heavy metal concentrations of bryophytes in the northeastern United States. *Oikos* **1976**, *27*, 445–456. [CrossRef]

13. Percy, K.E. Heavy metal and sulphur concentrations in *Sphagnum magellanicum* Brid. in the maritime provinces, Canada. *Water Air Soil Pollut.* **1982**, *19*, 341–349.

14. Rühling, Å.; Steinnes, E. *Atmospheric Heavy Metal Deposition in Europe 1995–1996*; Nordic Council of Ministers: Copenhagen, Denmark, 1998.

15. Wolterbeek, H.T. Biomonitoring of trace element air pollution: Principles, possibilities and perspectives. *Environ. Pollut.* **2002**, *120*, 11–21. [CrossRef]

16. Lee, C.S.L.; Li, X.; Zhang, G.; Peng, X.; Zhang, L. Biomonitoring of trace metals in the atmosphere using moss (*Hypnum plumaeforme*) in the Nanling Mountains and the Pearl River Delta, Southern China. *Atmos. Environ.* **2005**, *39*, 397–407. [CrossRef]

17. Elias, C.; Fernandes, E.A.N.; França, E.J.; Bacchi, M.A.; Tagliaferro, F.S. Native bromeliads as biomonitors of airborne chemical elements in a Brazilian resting forest. *J. Radioanal. Nucl. Chem.* **2008**, *278*, 423–427. [CrossRef]

18. Harmens, H.; Norris, D.A.; Steinnes, E.; Kubin, E.; Piispanen, J.; Alber, R.; Aleksiayenak, Y.; Blum, O.; Coşkun, M.; Dam, M.; et al. Mosses as biomonitors of atmospheric heavy metal deposition: Spatial and temporal trends in Europe. *Environ. Pollut.* **2010**, *158*, 3144–3156. [CrossRef]

19. Harmens, H.; Norris, D.A.; Sharps, K.; Mills, G.; Alber, R.; Aleksiayenak, Y.; Blum, O.; Cucu-Man, S.M.; Dam, M.; De Temmerman, L.; et al. Heavy metal and nitrogen concentrations in mosses are declining across Europe whilst some "hotspots" remain in 2010. *Environ. Pollut.* **2015**, *200*, 93–104. [CrossRef]

20. Harmens, H.; Buse, A.; Büker, P.; Norris, D.; Mills, G.; Williams, B. Heavy metal concentrations in European mosses: 2000/2001 survey. *J. Atmos. Chem.* **2004**, *49*, 425–436. [CrossRef]

21. Steinnes, E.; Uggerud, H.T.; Pfaffhuber, K.A.; Berg, T. *Atmospheric Deposition of Heavy Metals in Norway—National Moss Survey 2015. M–595*; Norwegian Institute for Air Research: Trondheim, Norway, 2016.

22. Barandovski, L.; Frontasyeva, M.V.; Stafilov, T.; Šajn, R.; Ostrovnaya, T.M. Multi-element atmospheric deposition in Macedonia studied by the moss biomonitoring technique. *Environ. Sci. Pollut. Res.* **2015**, *22*, 16077–16097. [CrossRef]

23. Mahapatra, B.; Dhal, N.K.; Dash, A.K.; Panda, B.P.; Panigrahi, K.C.S.; Pradhan, A. Perspective of mitigating atmospheric heavy metal pollution: Using mosses as biomonitoring and indicator organism. *Environ. Sci. Pollut. Res.* **2019**, *26*, 29620–29638. [CrossRef]

24. Stafilov, T.; Šajn, R.; Pančevski, Z.; Boev, B.; Frontasyeva, M.V.; Strelkova, L.P. Heavy metal contamination of surface soils around a lead and zinc smelter in the Republic of Macedonia. *J. Hazard. Mater.* **2010**, *175*, 896–914. [CrossRef] [PubMed]

25. Stafilov, T.; Šajn, R.; Boev, B.; Cvetković, J.; Mukaetov, D.; Andreevski, M.; Lepitkova, S. Distribution of some elements in surface soil over the Kavadarci Region, Republic of Macedonia. *Environ. Earth Sci.* **2010**, *61*, 1515–1530. [CrossRef]
26. Stafilov, T.; Aliu, M.; Šajn, R. Arsenic in surface soils affected by mining and metallurgical processing in K. Mitrovica Region, Kosovo. *Int. J. Environ. Res. Public Health* **2010**, *7*, 4050–4061. [CrossRef] [PubMed]
27. Balabanova, B.; Stafilov, T.; Bačeva, K.; Šajn, R. Biomonitoring of atmospheric pollution with heavy metals in the copper mine vicinity located near Radoviš. Republic of Macedonia. *J. Environ. Sci. Health A* **2010**, *45*, 1504–1518. [CrossRef]
28. Balabanova, B.; Stafilov, T.; Šajn, R.; Bačeva, K. Distribution of chemical elements in attic dust as reflection of lithology and anthropogenic influence in the vicinity of copper mine and flotation. *Arch. Environ. Contam. Toxicol.* **2011**, *6*, 173–184. [CrossRef]
29. Bačeva, K.; Stafilov, T.; Šajn, R.; Tănăselia, C.; Iić Popov, S. Distribution of chemical elements in attic dust in the vicinity of ferronickel smelter plant. *Fresenius Environ. Bull.* **2011**, *20*, 2306–2314.
30. Bačeva, K.; Stafilov, T.; Šajn, R.; Tănăselia, C. Moss biomonitoring of air pollution with heavy metals in the vicinity of a ferronickel smelter plant. *J. Environ. Sci. Health A* **2012**, *47*, 645–656. [CrossRef]
31. Balabanova, B.; Stafilov, T.; Šajn, R.; Bačeva, K. Characterisation of heavy metals in lichen species *Hypogymnia physodes* and *Evernia prunastri* due to biomonitoring of air pollution in the vicinity of copper mine. *Int. J. Environ. Res. Public Health* **2012**, *6*, 779–792.
32. Barandovski, L.; Frontasyeva, M.V.; Stafilov, T.; Šajn, R.; Pavlov, S.; Enimiteva, V. Trends of atmospheric deposition of trace elements in Macedonia by the moss biomonitoring technique. *J. Environ. Sci. Health A* **2012**, *47*, 2000–2015. [CrossRef]
33. Barandovski, L.; Stafilov, T.; Šajn, R.; Frontasyeva, M.V.; Bačeva, K. Air pollution study in Macedonia using a moss biomonitoring technique, ICP-AES, and AAS. *Maced. J. Chem. Chem. Eng.* **2013**, *32*, 89–107. [CrossRef]
34. Boev, B.; Stafilov, T.; Bačeva, K.; Šorša, A.; Boev, I. Influence of a nickel smelter plant on the mineralogical composition of attic dust in the Tikveš Valley, Republic of Macedonia. *Environ. Sci. Pollut. Res.* **2013**, *20*, 3781–3788. [CrossRef] [PubMed]
35. Bačeva, K.; Stafilov, T.; Šajn, R.; Tănăselia, C. Air dispersion of heavy metals in vicinity of the As-Sb-Tl abounded mine and responsiveness of moss as a biomonitoring media in small scale investigations. *Environ. Sci. Pollut. Res.* **2013**, *20*, 8763–8779. [CrossRef] [PubMed]
36. Stafilov, T.; Šajn, R.; Alijagić, J. Distribution of arsenic, antimony, and thallium in soil in Kavadarci and its surroundings, Republic of Macedonia. *Soil Sediment Contam.* **2013**, *22*, 105–118. [CrossRef]
37. Stafilov, T. Environmental pollution with heavy metals in the Republic of Macedonia. *Contrib. Sect. Nat. Math. Biotechnol. Sci. MASA* **2014**, *35*, 81–119. [CrossRef]
38. Bačeva, K.; Stafilov, T.; Matevski, V. Bioaccumulation of heavy metals by endemic viola species from the soil in the vicinity of the As-Sb-Tl mine "Allchar", Republic of Macedonia. *Int. J. Phytoremediat.* **2014**, *16*, 347–365. [CrossRef]
39. Balabanova, B.; Stafilov, T.; Šajn, R. Variability assessment for multivariate analysis of elemental distributions due to anthropogenic and geogenic impact in lead-zinc mine and flotation environ (moss sampling media). *Geol. Maced.* **2014**, *28*, 35–41.
40. Balabanova, B.; Stafilov, T.; Šajn, R.; Bačeva, K. Comparison of response of moss, lichens and attic dust to geology and atmospheric pollution from copper mine. *Int. J. Environ. Sci. Technol.* **2014**, *11*, 517–528. [CrossRef]
41. Angelovska, S.; Stafilov, T.; Šajn, R.; Bačeva, K.; Balabanova, B. Moss biomonitoring of air pollution with heavy metals in the vicinity of Pb-Zn mine "Toranica" near the town of Kriva Palanka. *Modern Chem. Appl.* **2014**, *2*, 1–6.
42. Balabanova, B.; Stafilov, T.; Šajn, R.; Tănăselia, C. Multivariate extraction of dominant geochemical markers for 69 elements deposition in Bregalnica River Basin, Republic of Macedonia (moss biomonitoring). *Environ. Sci. Pollut. Res.* **2016**, *23*, 22852–22870. [CrossRef]
43. Balabanova, B.; Stafilov, T.; Šajn, R.; Bačeva Andonovska, K. Quantitative assessment of metal elements using moss species as biomonitors in downwind area of lead-zinc mine. *J. Environ. Sci. Health A* **2017**, *52*, 290–301. [CrossRef]
44. Stafilov, T.; Šajn, R. *Geochemical Atlas of the Republic of Macedonia*; Faculty of Natural Sciences and Mathematics, Sc. Cyril and Methodius University: Skopje, North Macedonia, 2016.

45. Stafilov, T.; Šajn, R.; Arapčeska, M.; Kungulovski, I.; Alijagić, J. Geochemical properties of topsoil around the coal mine and thermoelectric power plant. *J. Environ. Sci. Health A* **2018**, *53*, 793–808. [CrossRef] [PubMed]

46. Stafilov, T.; Špirić, Z.; Glad, M.; Barandovski, L.; Bačeva Andonovska, K.; Šajn, R.; Antonić, O. Study of nitrogen pollution in the Republic of North Macedonia by moss biomonitoring and Kjeldahl method. *J. Environ. Sci. Health A* **2020**, *55*, 759–764. [CrossRef] [PubMed]

47. Bačeva, K.; Stafilov, T.; Šajn, R. Biomonitoring of nickel air pollution near the city of Kavadraci, Republic of Macedonia. *Ecol. Protect. Environ.* **2009**, *12*, 57–69.

48. Dimovska, B.; Stafilov, T.; Šajn, R.; Tănăselia, C. Moss biomonitoring of air pollution with arsenic in Bitola and the Environs, Republic of Macedonia. *Geol. Maced.* **2013**, *27*, 5–11.

49. Dimovska, B.; Stafilov, T.; Šajn, R.; Bačeva, K. Distribution of lead and zinc in soil over the Bitola region, Republic of Macedonia. *Geol. Maced.* **2014**, *28*, 87–91.

50. Stafilov, T.; Šajn, R.; Sulejmani, F.; Bačeva, K. Geochemical properties of topsoil around the open coal mine and Oslomej thermoelectric power plant R.; Macedonia. *Geol. Croat.* **2014**, *67*, 33–44. [CrossRef]

51. Blouet, B.W. *The EU and Neighbors: A Geography of Europe in Modern World*; John Wiley & Sons: New York, NY, USA, 2008.

52. Environmental Statistics, State Statistical Office of the Republic of Macedonia, Skopje, 2019. Available online: http://www.stat.gov.mk/Publikacii/ZivotnaSredina2019.pdf (accessed on 18 August 2020).

53. Lazarevski, A. *Climate in Macedonia*; Kultura: Skopje, North Macedonia, 1993. (In Macedonian)

54. Balabanova, B.; Stafilov, T.; Šajn, R. Enchasing anthropogenic element trackers for evidence of long-term atmospheric depositions in mine environ. *J. Environ. Sci. Health A* **2019**, *54*, 988–998. [CrossRef]

55. Tomovski, D.; Bačeva Andonovska, K.; Šajn, R.; Karadjov, M.; Stafilov, T. Distribution of chemical elements in sediments and alluvial soils from the Crna Reka river basin. *Geol. Maced.* **2019**, *33*, 125–145.

56. Vasilevska, S.; Stafilov, T.; Šajn, R. Distribution of trace elements in sediments and soil from Crn Drim River Basin, Republic of Macedonia. *SN Appl. Sci.* **2019**, *1*, 1–16. [CrossRef]

57. Stafilov, T.; Šajn, R.; Ahmeti, L. Geochemical characteristics of soil of the city of Skopje, Republic of Macedonia. *J. Environ. Sci. Health A* **2019**, *54*, 972–987. [CrossRef]

58. Stafilov, T.; Šajn, R.; Mickovska, I. Distribution of chemical elements in soil from Crn Drim River basin, Republic of Macedonia. *Contrib. Sect. Nat. Math. Biotech. Sci. MASA* **2019**, *40*, 73–92. [CrossRef]

59. Stafilov, T.; Šajn, R. Spatial distribution and pollution assessment of heavy metals in soil from the Republic of North Macedonia. *J. Environ. Sci. Health A* **2019**, *54*, 1457–1474. [CrossRef] [PubMed]

60. Dumurdzanov, N.; Serafimovski, T.; Burchfiel, B.C. Cenozoic tectonics of Macedonia and its relation to the South Balkan extensional. *Geosphere* **2005**, *1*, 1–22. [CrossRef]

61. Jovanovski, G.; Boev, B.; Makreski, P. *Minerals from the Republic of Macedonia with an Introduction to Mineralogy*; Macedonian Academy of Sciences and Arts: Skopje, North Macedonia, 2012.

62. ICP Vegetation Coordination Centre. *ICP Vegetation Experimental Protocol for the 2001 Season*; ICP Vegetation Coordination Centre: Bangor, UK, 2001.

63. Harmens, H. (Ed.) *Heavy Metals in European Mosses: 2005/2006 Survey—Monitoring Manual*; International Cooperative Programme on Effects of Air Pollution on Natural Vegetation and Crops, Centre for Ecology and Hydrology: Bangor, UK, 2005.

64. Harmens, H. (Ed.) *Heavy Metals in European Mosses: 2010 Survey—Monitoring Manual*; International Cooperative Programme on Effects of Air Pollution on Natural Vegetation and Crops, Centre for Ecology and Hydrology: Bangor, UK, 2010.

65. Snedecor, G.W.; Cochran, W.G. *Statistical Methods*; The Iowa State University Press: Ames, IA, USA, 1967.

66. Hollander, M.; Wolfe, D.A. *Nonparametric Statistical Methods*, 2nd ed.; Wiley: New York, NY, USA, 1999.

67. Box, G.E.P.; Cox, D.R. An analysis of transformations. *J. R. Stat. Soc. Ser. B* **1962**, *26*, 211–252. [CrossRef]

68. Shoji, T. Enrichment ratio-tonnage diagrams for resource assessment. *Nat. Resour. Res.* **2002**, *11*, 273–287. [CrossRef]

69. La Maitre, R.W. *Numerical Petrology, Statistical Interpretation of Geochemical Data*; Elsevier: Amsterdam, The Netherlands, 1982; p. 281.

70. Davis, J.C. *Statistic and Data Analysis in Geology*; Willey: New York, NY, USA, 1986; p. 646.

71. Reimann, C.; Filzmoser, P.; Garrett, R.G. Factor analysis applied to regional geochemical data: Problems and possibilities. *Appl. Geochem.* **2002**, *17*, 185–206. [CrossRef]

72. Templ, M.; Filzmoser, P.; Reimann, C. Cluster analysis applied to regional geochemical data: Problems and possibilities. *Appl. Geochem.* **2008**, *23*, 2198–2213. [CrossRef]
73. Filzmoser, P.; Garrett, R.G.; Reimann, C. Multivariate outlier detection in exploration geochemistry. *Comput. Geosci.* **2005**, *31*, 579–587. [CrossRef]
74. Frontasyeva, M.V.; Steinnes, E. Marine gradients of halogens in moss studied by epithermal neutron activation analysis. *J. Radioanal. Nucl. Chem.* **2004**, *261*, 101–106. [CrossRef]
75. Čeburnis, D.; Steinnes, E.; Kveitkus, K. Estimation of metal uptake efficiencies from precipitation in mosses in Lithuania. *Chemosphere* **1999**, *38*, 445–455. [CrossRef]
76. State Statistical Office of the Republic of Macedonia. *Environmental Statistics*; State Statistical Office of the Republic of Macedonia: Skopje, North Macedonia, 2011.
77. State Statistical Office of the Republic of Macedonia. *Environmental Statistics*; State Statistical Office of the Republic of Macedonia: Skopje, North Macedonia, 2013.
78. De Agostini, A.; Cortis, P.; Cogoni, A. Monitoring of air pollution by moss bags around an oil refinery: A critical evaluation over 16 years. *Atmpshere* **2020**, *11*, 272. [CrossRef]
79. Pais, I.; Jones, J.B. *The Handbook of Trace Elements*; St. Lucie Press: Boca Raton, FL, USA, 1997.
80. Stafilov, T.; Šajn, R.; Balabanova, B.; Bačeva, K. Distribution of heavy metals in attic and deposited dust in the vicinity of copper ore processing and ferronickel smelter plants in the Republic of Macedonia. In *Dust: Sources, Environmental Concerns and Control*; Wouters, L.B., Pauwels, M., Eds.; Nova Science Publishers: Hauppauge, NY, USA, 2012; pp. 57–98.
81. Stafilov, T.; Šajn, R.; Boev, B.; Cvetković, J.; Mukaetov, D.; Andreevski, M. *Geochemical Atlas of Kavadarci and the Environs*; Faculty of Natural Sciences and Mathematics: Skopje, North Macedonia, 2008.
82. Stafilov, T.; Peeva, L.; Nikov, B.; de Koning, A. Industrial hazardous waste in the Republic of Macedonia. In *Applied Environmental Geochemistry—Anthropogenic Impact on Human Environment in the SE Europe*; Šajn, R., Žilbert, G., Alijagić, J., Eds.; Geological Survey of Slovenia: Ljubljana, Slovenia, 2009; pp. 108–112.
83. Steinnes, E.; Frontasyeva, M.V. Marine gradients of halogens in soil studied by epithermal neutron activation analysis. *J. Radioanal. Nucl. Chem.* **2002**, *253*, 173–177. [CrossRef]

Article

Mosses as Bioindicators of Heavy Metal Air Pollution in the Lockdown Period Adopted to Cope with the COVID-19 Pandemic

Nikita Yushin [1], Omari Chaligava [1,2], Inga Zinicovscaia [1,3,*]**, Konstantin Vergel [1] and Dmitrii Grozdov [1]**

[1] Joint Institute for Nuclear Research, Joliot-Curie 6, 141980 Dubna, Russia; ynik_62@mail.ru (N.Y.); omar.chaligava@ens.tsu.edu.ge (O.C.); verkn@mail.ru (K.V.); dsgrozdov@rambler.ru (D.G.)

[2] Faculty of Exact and Natural Science, Ivane Javakhishvili Tbilisi State University, Chavchavadze ave. 3, 0179 Tbilisi, Georgia

[3] Horia Hulubei National Institute for R&D in Physics and Nuclear Engineering, 30 Reactorului, MG-6 Bucharest-Magurele, Romania

* Correspondence: zinikovskaia@mail.ru; Tel.: +7-496-216-5609

Received: 9 October 2020; Accepted: 1 November 2020; Published: 3 November 2020

Abstract: The coronavirus disease, COVID-19, has had a great negative impact on human health and economies all over the world. To prevent the spread of infection in many countries, including the Russian Federation, public life was restricted. To assess the impact of the taken actions on air quality in the Moscow region, in June 2020, mosses *Pleurosium shreberi* were collected at 19 sites considered as polluted in the territory of the region based on the results of the previous moss surveys. The content of Cd, Cr, Cu, Fe, Ni, and Pb in the moss samples was determined using atomic absorption spectrometry. The obtained values were compared with the data from the moss survey performed in June 2019 at the same sampling sites. Compared to 2019 data, the Cd content in moss samples decreased by 2–46%, while the iron content increased by 3–127%. The content of Cu, Ni, and Pb in mosses decreased at most sampling sites, except for the eastern part of the Moscow region, where a considerable number of engineering and metal processing plants operate. The stay-at-home order issued in the Moscow region resulted in a reduction of vehicle emissions affecting air quality, while the negative impact of the industrial sector remained at the level of 2019 or even increased.

Keywords: COVID-19; air pollution; metals; industry; moss survey; biomonitoring

1. Introduction

On March 11, 2020, the World Health Organization declared COVID-19 as a global pandemic [1]. On March 12, based on a decree of the Government of the Moscow region, a self-isolation regime was introduced. According to the official decision, only essential services such as healthcare, logistics, food supply, public transport, and industrial enterprises due to technical reasons did not cease to operate. A significant part of the regular national and international flights and train services was cancelled. Since the majority of the population switched to remote working, the number of vehicles has dropped significantly.

New rules adopted in many countries to help slow the spread of COVID-19 resulted in a decrease in the negative impact on the environment in some regions of the world [2]. Nadzir et al. [3] measured the concentrations of CO, $PM_{2.5}$, and PM_{10} in Malaysia and observed that the concentration of pollutants declined significantly, by ≈20 to 60%, during the Control Order in Malaysia (MCO) days at most studied locations. At the same time, in Kota Damansara, the level of pollutants significantly increased due to local anthropogenic activity. COVID-19 resulted in a worldwide decrease in the concentration of CO_2

in March 2020 by 7% in comparison with the monthly concentration in 2019 [4]. In three Chinese cities (Chongqing, Luzhou, and Chengdu) concentrations of $PM_{2.5}$, PM_{10}, SO_2, CO, and NO_2 in February 2020 were lower by 17.9%–62.1% than the values determined in February 2017–2019 [5]. In three other cities in China (Wuhan, Jingmen, and Enshi) concentrations of the abovementioned pollutants measured in February 2020 declined by 27.9%–61.4% compared to February 2017–2019. At the same time, in the period of January–March 2020, an increase of O_3 concentration of up to 15% was noticed [6]. In Turkey, the concentrations of PM_{10} and NO_2 in 2016 and 2020 were compared, and no significant difference in PM_{10} concentrations was observed. In the after-lockdown period, the NO_2 concentrations were lowered by 11.8% [7]. A decrease in the $PM_{2.5}$ concentration in Wuhan, Daegu, and Tokyo by 29.9%, 20.9%, and 3.6%, respectively, took place after one month of COVID-related restrictions. The concentration of NO_2 also declined in all three cities, with the most pronounced decrease being by 53.2% in Wuhan [2]. In Brazil, a remarkable decrease in CO and NO_2 concentrations was observed during the lockdown period [8].

PM contains various metals, including some elements recognized as human carcinogens. Sources of metals in the atmosphere can be natural: soil dust, volcano eruptions, and forest fires, or they can be anthropogenic, which include road traffic, industry, and thermal power plants [9]. It is often difficult to conduct measurements of concentrations of atmospheric metals over large territories. Passive moss biomonitoring proved to be a suitable technique to study the spatial distribution of metals over the large territories [10]. This technique is well recognized and widely applied in many European countries [11]. Although data obtained using moss biomonitors do not correspond to the direct quantitative measurement of metal deposition [12], this information is very useful for regions where it is difficult to obtain official data related to air pollution. In the Moscow region, moss survey studies were performed in 2004 [13], 2014 [14], and 2019 (data not yet published).

The aim of the present study was to find out whether moss can be used as a tool to assess the impact of the restrictions due to the lockdown imposed to face the pandemic emergency on air quality in a relatively short time (2.5 mounts). For this purpose, the data obtained for moss samples collected in the Moscow region in June 2020 were compared with the data obtained in 2019 for the same collection sites.

2. Experiments

2.1. Study Area and Sampling

The Moscow region, which is a subject of the Russian Federation, is one of the most densely populated and industrially developed regions of the country [14]. The Moscow region can be considered the economic core of the country, with about 70% of Russia's financial wealth concentrated here. It is located in the center of railway and air networks, reaching Transcaucasia, Central Asia, and the Pacific, and it is connected by rivers and canals to the Baltic, White, Azov, Black, and Caspian Seas [15]. In the center of the region is Moscow, the capital of the Russian Federation, and the most important industrial city, transportation hub, educational and cultural center of the region [13,15]. The industrial sector of the Moscow region includes metallurgical, oil refining, mechanical engineering, food, energy, and chemical enterprises [14]. The main industrial centers in the Moscow region are: Krasnogorsk, Lyubertsy, Mytishchi, Klin, Noginsk, Pavlovsky Posad, Voskresensk, Kolomna, Dmitrov, Klin, Elektrostal, Balashikha, and Sergiyev Posad.

In 2019 and 2020, moss *Pleurosium shreberi* was collected on the sheared tree bark. Moss sampling was performed in June 2019 and June 2020 at 19 sampling sites (Figure 1) in accordance with [16]. The selection of the sampling sites in 2020 was based on the moss survey data obtained in 2019 (analyzed but not yet published). Moss samples were collected at sites with the highest metal concentrations in 2019. After collection, samples were cleaned of soil particles and other contaminants. The upper 3–4 cm of green and green-brown shoots from the top of the moss were separated and dried at 105 °C to constant weight. For analysis, approximately 0.2 g of moss was placed in a Teflon vessel and treated with 2 mL of concentrated nitric acid and 1 mL of hydrogen peroxide. The Teflon vessels were put into a microwave digestion system (Mars; CEM, USA) for complete digestion. Digestion was performed in two steps: (1) ramp: temperature 180 °C, time 15 min, power 400 W, and pressure 20 bar; (2) hold: temperature 160 °C, time 10 min, power 400 W, and pressure 20 bar. Digests were quantitatively transferred to 100-mL calibrated flasks and made up to the volume with bidistilled water.

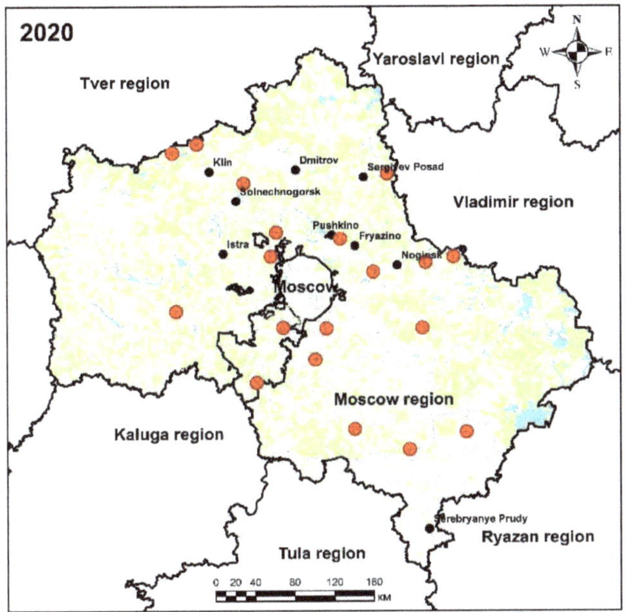

Figure 1. Sampling map (in 2019 and 2020 samples were collected in the same sampling sites).

2.2. Chemical Analysis

The content of Cd, Cu, Pb, Cr, Ni, Fe, V, and Sb in the moss samples was determined by means of atomic absorption spectrometry using a Thermo Scientific™ iCE™ 3400 AA spectrometer (Thermo Scientific, Waltham, MA, USA) with electrothermal (graphite furnace) atomization. Stock solutions (AAS standard solution; Merck, Germany) with metal concentrations of 1 g/L were used to prepare calibration solutions. The National Institute of Standards and Technology (NIST) reference materials 1570a (Trace Elements in Spinach Leaves) and 1575a (Pine Needles) were used to ensure the quality control of measurements. The difference between determined and certified values did not exceed 5%.

2.3. Data Processing

Statistical analysis of the data was done using Excel 2016 (Microsoft, Redmond, Washington, USA) and IBM SPSS software (IBM, Armonk, New York, USA). The Wilcoxon signed-rank test [17] was applied to define differences between the values obtained in 2019 and 2020. The ArcGis 10.6 software (Esri, Redlands, California, USA) was used to build maps showing the spatial distribution of elements.

To quantify the anthropogenic influence on the environment, several indices were calculated, such as the contamination factor (CF), the Geo-accumulation Index (I_{geo}), and the pollution load index (PLI).

The contamination factor CF is defined as:

$$CF = \frac{C_m}{C_b} \qquad (1)$$

where C_m is the measured content of the metal at any given site and C_b is the background level for that metal [18].

CF <1 no contamination; 1–2 suspected; 2–3.5 slight; 3.5–8 moderate; 8–27 severe; and >27 extreme [19].

The index of geo-accumulation, I_{geo}, was calculated using the following formula:

$$C_F = \frac{C_M}{1.5 C_B} \tag{2}$$

where $\frac{C_M}{C_B}$ is the contamination factor. The factor of 1.5 is introduced to minimize the effect of possible variations in the background [20].

I_{geo} <0 no contamination; 0–1 slightly polluted; 1–2 moderately polluted; 2–3 moderately to severely polluted; 3–4 severely polluted; 4–5 severely to extremely polluted; and I_{geo} > 5 extremely polluted [21].

The PLI represents the nth order geometric mean of the entire set of CF regarding the contaminating elements as follows:

$$PLI = \sqrt[n]{\prod_{i=1}^{n} C_{F,i}}, \tag{3}$$

where n is the total number of contaminating elements.

PLI < 1 (non polluted); 1 ≤ PLI < 2 (slight polluted); 2 ≤ PLI < 3 (moderately polluted); PLI < 3 (highly polluted) [22].

3. Results and Discussion

Eight elements were determined in the analyzed moss samples using atomic absorption spectrometry (AAS). Since the concentrations of V and Sb were below the detection limits, these elements were excluded from further discussion. The results of the statistical analysis for the analyzed elements in 2019 and 2020 are presented in Table 1.

Table 1. Descriptive statistics of results for moss samples collected in 2019 and 2020 (in mg/kg dry weight (d.w.)).

Element	Year	Range	Md	Mean ± St. Dev	Q1	Q3	CV (%)	p
Cd	2019	0.11–0.64	0.30	0.34 ± 0.14	0.23	0.41	41.9	<0.05
	2020	0.14–0.52	0.24	0.26 ± 0.09	0.22	0.30	33.9	
Pb	2019	1.71–17.2	4.41	5.62 ± 3.45	3.80	7.02	61.4	>0.05
	2020	1.79–13.6	4.60	5.31 ± 3.14	3.21	6.26	59.2	
Cu	2019	6.38–21	8.98	10.3 ± 3.93	7.29	12.9	38.3	<0.05
	2020	4.72–15.8	7.76	8.8 ± 3.18	6.64	9.36	36.3	
Cr	2019	1.1–3.09	1.98	1.87 ± 0.58	1.32	2.16	30.8	>0.05
	2020	1.01–4.29	1.98	2 ± 0.75	1.46	2.18	37.6	
Ni	2019	1.53–5.86	2.91	3.36 ± 1.17	2.67	4.27	34.9	<0.05
	2020	2.59–7.35	3.85	4.27 ± 1.27	3.37	4.61	29.8	
Fe	2019	343–1175	579	607 ± 237	446	682	39.1	<0.05
	2020	309–2551	693	846 ± 476	594	933	56.2	

Md: median; P90–90 percentile; St. Dev.: standard deviation; CV: coefficient of variance, *p*-values for differences were obtained from Wilcoxon signed-rank test.

According to the Wilcoxon test, no significant differences ($p > 0.05$) were found for Pb and Cr, while for Cu, Cd, Fe, and Ni, significant differences ($p < 0.05$) between the median concentrations were revealed. The CV values for all elements in both years were less than 75%, which points to the main influence of the regional source of pollution [14].

The median values determined in the present study were compared with the data obtained for the previous moss surveys performed in the Moscow region in 2004 and 2014 (Table 2). The content of Cd, Cu, and Ni in moss samples was almost at the same level in the period 2004–2020. The higher values of the mean Pb content in 2019–2020 in comparison with 2014 can be explained by the fact that in 2019–2020 and 2014, the moss samples were collected at different sampling sites: in 2014—in the north-eastern part of Moscow, whereas in 2019–2020—in places considered to be potentially highly polluted. The iron content was the highest in 2014, followed by 2004, 2020, and 2019, while the content of Cr in mosses collected in 2019 and 2020 was significantly lower than in 2014 and 2004.

Table 2. Comparison between the median values obtained in the present study and data reported for previous surveys (in mg/kg d.w.).

Element	Moscow Region 2020	Moscow Region 2019	Moscow Region 2014 [14]	Moscow Region 2004 [13]
Cd	0.24	0.30	0.3	x
Pb	4.60	4.41	0.67	x
Cu	7.76	8.98	7.1	x
Cr	1.98	1.98	3.2	3.1
Ni	3.85	2.91	3.2	2.4
Fe	693	579	1050	800

In order to reveal the differences in metal uptake by moss in 2019 and 2020, the element distribution maps of are given in Figure 2. Natural sources of Cd emissions are volcanic activity and release by vegetation [23]. The main anthropogenic sources are non-ferrous metal production, waste incineration, dust generated during the operation of vehicles, and resuspension of road dust [23,24]. Hjortenkrans et al. [25] showed that emissions from brake linings/tire tread rubber contain a wide range of elements, including Cd, Cu, Pb, Sb, and Zn. Comparing the data obtained in 1998 and 2005, the authors showed that Cu and Zn emissions remained unchanged in the studied period, suggesting that brake lining is one of the main sources of these metals. On the other hand, a pronounced decrease in the content of Pb and Cd in the studied period was observed. The input of anthropogenic sources of Cd emission significantly exceeded the release from natural resources. According to the data obtained in 2020, the content of Cd in the Moscow region diminished by 2–46%. The most pronounced decrease was noticed near the cities of Sergeyev Posad, Solnechnogorsk, and Domodedovo. According to the reports by the national authorities, during the period of self-isolation, traffic in the Moscow region declined by 50%.

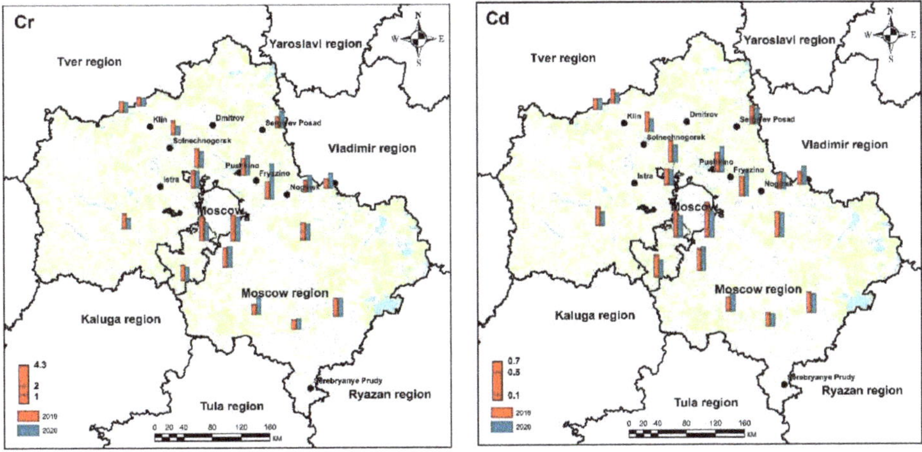

Figure 2. *Cont.*

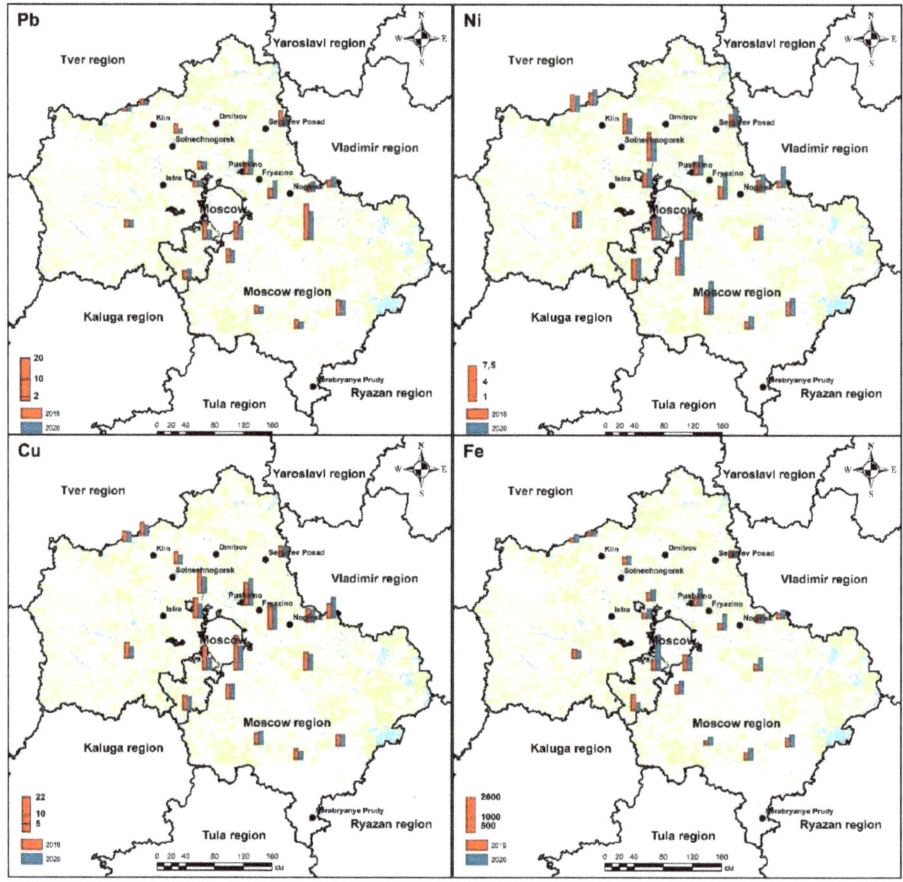

Figure 2. Maps of element content in moss samples collected in 2019 and 2020 in the Moscow region (in mg/kg d.w).

The increase of Cd concentrations near Klin and Stupino can be explained by the impact of metallurgical and engineering complexes.

The major sources of Cr in the environment are metal processing, coal burning, and vehicles. Chromium is one of the most abundant metals in diesel particles [26]. Compared to 2019, in 2020, the content of Cr in moss samples decreased by 7–35%, mainly in the north-eastern part of the Moscow region. In Sergeyev Posad, the increase in Cr content by 59% may be related to the activities of an engineering plant and the production of paint and coatings. The rise of Cr content in satellite cities near Moscow by 42–100% may be associated with the industrial activity of machine-building, metallurgical, and chemical plants, which were operating at full capacity during the self-isolation period.

Brake wear emissions account for up to 75% of Cu emissions into the air [27]. A decrease in the Cu content in moss samples collected in 2020 by 8–50% compared to the data for 2019 confirms this hypothesis. At the same time, an increase in its content by 38% in satellite cities near Moscow points to the dominant contribution of industrial activity to Cu emissions.

The main sources of Fe in the atmosphere are industrial and metallurgical processes, combustion of fossil fuels, transport, as well as resuspension of crustal materials and road dust [28]. In contrast to the previously discussed element, the content of Fe in the moss samples collected in 2020 increased significantly in comparison with the data for 2019. This increase of 3–217% can be attributed mainly to the resuspension of crustal materials.

Nickel can be released into the atmosphere from natural and anthropogenic sources. The main anthropogenic sources are fossil fuel combustion, smelting of ferrous and non-ferrous metals, waste incineration, and other various sources [29]. The increase in Ni content in satellite cities around Moscow in 2020 may be associated with industrial activities of metallurgical plants in Electrostal, Shchyolkovo, and Podolsk, and engineering plants in Electrostal and Podolsk. The rise of Ni content at other sampling sites is mainly due to vehicles and resuspension of dust particles. In spite of the fact that Pb content has been declining in many countries due to the introduction of unleaded fuels, vehicles continue to be one of the main sources of Pb emissions. On a regional scale, industrial activities may contribute to emissions of lead into the air [30]. A significant decrease in Pb content was noticed in 2020 in comparison with 2019 (up to 48%), indicating a decline in traffic flow. A considerable increase in Pb content by 30–65% in the western part of the Moscow region indicates the dominant role of industrial activity in Pb emission in this area.

The strength of association of the chemical elements in the moss samples collected in 2019 and 2020 can be seen in Table 3. The Spearman correlation coefficient between 0.5 and 0.7 indicated a good association between the elements, whereas r in the range of 0.7–1.0 shows strong association of elements [31].

Table 3. Spearman correlation coefficient between element content in mosses collected in Moscow region in 2019 and 2020.

		Cd	Pb	Cu	Cr	Ni	Fe
2019	Cd	**1.00**					
	Pb	**0.76 ****	**1.00**				
	Cu	0.25	0.38	1.00			
	Cr	0.43	**0.60 ****	**0.71 ****	1.00		
	Ni	0.28	0.14	0.28	**0.55 ***	**1.00**	
	Fe	**0.56 ***	**0.48 ***	**0.46 ***	**0.66 ****	0.36	**1.00**
		Cd	Pb	Cu	Cr	Ni	Fe
2020	Cd	**1.00**					
	Pb	**0.55 ***	**1.00**				
	Cu	0.36	**0.74 ****	**1.00**			
	Cr	0.41	**0.67 ****	**0.66 ****	1.00		
	Ni	0.42	0.39	**0.61 ****	**0.73 ****	**1.00**	
	Fe	0.22	**0.69 ****	**0.61 ****	**0.71 ****	0.40	**1.00**

* Correlation is significant at the 0.05 level (2-tailed). ** Correlation is significant at the 0.01 level (2-tailed).

In 2019, a high positive correlation in Pb-Cd and Cr-Cu pairs was observed. Road traffic may be responsible for Cr, Cu, Pb, V, and Zn emissions, whereas fuel combustion is the major source of Cr, Cu, and V [29]. Good positive correlations were found between Fe-Cd, Cr-Pb, Ni-Cr, and Fe-Cr pairs of elements, and this may be related to industrial activities. In 2020, a high correlation between Pb and Cd was observed again. A high positive correlation was also determined in the Cu-Pb, Ni-Cr, and Fe-Cr pairs. Good correlations were obtained between the Cr-Pb, Fe-Pb, Ni-Cu, and Fe-Cu pairs of elements. The obtained associations indicate a possible anthropogenic influence in the study area; however, their source could be a resuspension of soil particles.

The use of geochemical indices is important for the assessment of the contamination status of the investigated territory [32]. In the present study, the contamination factor (CF), Geo-accumulation Index (I_{geo}), and pollution load index (PLI) were calculated [14,32] (Table 4). In 2019, no contamination with Ni and moderate contamination with Cr, Fe, and Cu were determined. The CF values for Cd and Pb were higher than 3.0 and indicated considerable contamination. The situation changed in 2020 when CF pointed to moderate pollution for all elements. The decrease of CF values for Pb and Cd may be associated with the reduction of traffic flow, while the impact of the industrial activity remained at the same level. The I_{geo} results indicate that mosses were uncontaminated to moderately contaminated with the determined elements. In 2019, the moss samples were moderately contaminated with Cd, Pb, Cu, and Fe, and in 2020, with Cd and Pb.

Table 4. Mean values of the contamination factor CF and Geo-accumulation Index I_{geo} for the studied area.

Element \ Year	CF		I_{geo}	
	2019	2020	2019	2020
Cd	3.06	1.85	0.91	0.24
Pb	3.28	2.96	0.93	0.77
Cu	1.61	1.49	0.01	−0.09
Cr	1.39	1.58	−0.17	−0.02
Ni	0.92	1.30	−0.79	−0.26
Fe	1.74	1.48	0.12	−0.17

A PLI below 1.0 shows that elemental loads are approximately equal to the background level, and values above 1.0 indicate the degree of pollution [33]. As can be seen from the maps presented in Figure 3, in 2019, the entire territory under investigation belongs to the moderately polluted to unpolluted category (except for the sites near Troitsk and Domodedovo, which are moderately polluted). Similarly, in 2020, the sampling territory can be characterized as moderately polluted to unpolluted. The PLI values were higher than 2.0 near Domodedovo and Staraya Kupavna.

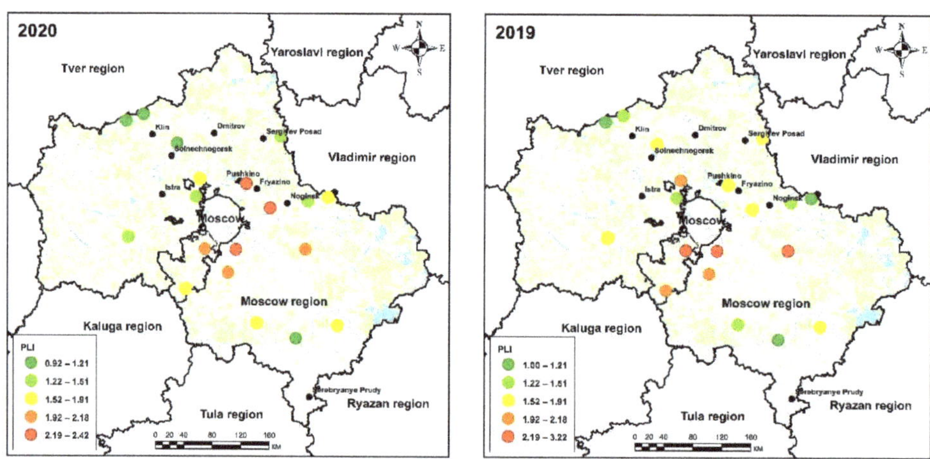

Figure 3. The map of the distribution of PLI values in Moscow region.

4. Conclusions

The results from two biomonitoring studies performed in the Moscow region in 2019 and 2020 were compared. Moss sampling has proven to be to be a suitable and low-cost indicator of heavy metal air pollution. The self-isolation period adopted to cope with the COVID-19 pandemic resulted in a decrease in Cd content in the Moscow region, while the content of other analyzed elements decreased in the north-eastern part of the Moscow region and remained the same or even increased in satellite cities near Moscow. Owing to the decline in the flow of traffic, stationary sources can be considered the primary source of metal emission into the atmosphere. In 2019 and 2020, according to the PLI values, the territory of the Moscow region was characterized as moderately polluted to unpolluted.

Author Contributions: Conceptualization, all authors; methodology, N.Y., I.Z., D.G., K.V., software, O.C.; formal analysis, all authors; writing—original draft preparation, I.Z.; writing—review and editing, all authors. All authors have read and agreed to the published version of the manuscript.

Funding: This research received no external funding.

Conflicts of Interest: The authors declare no conflict of interest.

References

1. WHO Archived: WHO Timeline—COVID-19. Available online: https://www.who.int/news-room/detail/27-04-2020-who-timeline---covid-19 (accessed on 23 July 2020).
2. Ma, C.J.; Kang, G.U. Air quality variation in wuhan, daegu, and tokyo during the explosive outbreak of covid-19 and its health effects. *Int. J. Environ. Res. Public Health* **2020**, *17*, 4119. [CrossRef]
3. Nadzir, M.S.M.; Ooi, M.C.G.; Alhasa, K.M.; Bakar, M.A.A.; Mohtar, A.A.A.; Nor, M.F.F.M.; Latif, M.T.; Hamid, H.H.A.; Ali, S.H.M.; Ariff, N.M.; et al. The impact of movement control order (MCO) during pandemic COVID-19 on local air quality in an urban area of Klang valley, Malaysia. *Aerosol Air Qual. Res.* **2020**, *20*, 1237–1248. [CrossRef]
4. Safarian, S.; Unnthorsson, R.; Richter, C. Effect of Coronavirus Disease 2019 on CO2 Emission in the World. *Aerosol Air Qual. Res.* **2020**, *20*, 1197–1203. [CrossRef]
5. Zhang, J.; Cui, K.; Wang, Y.-F.; Wu, J.-L.; Huang, W.-S.; Wan, S.; Xu, K. Temporal Variations in the Air Quality Index and the Impact of the COVID-19 Event on Air Quality in Western China. *Aerosol Air Qual. Res.* **2020**, *20*, 1552–1568. [CrossRef]
6. Xu, K.; Cui, K.; Young, L.H.; Hsieh, Y.K.; Wang, Y.F.; Zhang, J.; Wan, S. Impact of the COVID-19 event on air quality in central China. *Aerosol Air Qual. Res.* **2020**, *20*, 915–929. [CrossRef]
7. Kaskun, S. The effect of COVID-19 pandemic on air quality caused by tra c in Istanbul. *Res. Sq.* **2020**, 1–24. [CrossRef]
8. Siciliano, B.; Carvalho, G.; da Silva, C.M.; Arbilla, G. The Impact of COVID-19 Partial Lockdown on Primary Pollutant Concentrations in the Atmosphere of Rio de Janeiro and São Paulo Megacities (Brazil). *Bull. Environ. Contam. Toxicol.* **2020**, *105*, 2–8. [CrossRef]
9. Lequy, E.; Siemiatycki, J.; Leblond, S.; Meyer, C.; Zhivin, S.; Vienneau, D.; de Hoogh, K.; Goldberg, M.; Zins, M.; Jacquemin, B. Long-term exposure to atmospheric metals assessed by mosses and mortality in France. *Environ. Int.* **2019**, *129*, 145–153. [CrossRef]
10. Markert, B.A.; Breure, A.M.; Zechmeister, H.G. Chapter 1 Definitions, strategies and principles for bioindication/biomonitoring of the environment. *Trace Met. Contam. Environ.* **2003**, *6*, 3–39. [CrossRef]
11. Harmens, H.; Ilyin, I.; Mills, G.; Aboal, J.R.; Alber, R.; Blum, O.; Coşkun, M.; De Temmerman, L.; Fernández, J.A.; Figueira, R.; et al. Country-specific correlations across Europe between modelled atmospheric cadmium and lead deposition and concentrations in mosses. *Environ. Pollut.* **2012**, *166*, 1–9. [CrossRef]
12. Harmens, H.; Norris, D.A.; Steinnes, E.; Kubin, E.; Piispanen, J.; Alber, R.; Aleksiayenak, Y.; Blum, O.; Coşkun, M.; Dam, M.; et al. Mosses as biomonitors of atmospheric heavy metal deposition: Spatial patterns and temporal trends in Europe. *Environ. Pollut.* **2010**, *158*, 3144–3156. [CrossRef] [PubMed]
13. Vergel, K.N.; Frontasyeva, M.V.; Kamanina, I.Z.; Pavlov, S.S. Biomonitoring of heavy metals in the northeast of the Moscow region using moss-biomonitors. *Ecol. Urban. Territ.* **2009**, *3*, 88–95. (In Russian)
14. Vergel, K.; Zinicovscaia, I.; Yushin, N.; Frontasyeva, M.V. Heavy Metal Atmospheric Deposition Study in Moscow Region, Russia. *Bull. Environ. Contam. Toxicol.* **2019**, *103*, 435–440. [CrossRef]
15. Subregions, W.; Pulsipher, A. *World Regional Geography Atlas*; W.H. Freeman & Co.: New York, NY, USA, 2014.
16. Spranger, T. *Manual on Methodologies and Criteria for Modelling and Mapping Critical Loads & Levels and Air Pollution Effects, Risks and Trends*; Umweltbundesamt: Dessau-Roßlau, Germany, 2004.
17. Wilcoxon, F. Individual Comparisons by Ranking Methods. *Biom. Bull.* **1945**, *1*, 80. [CrossRef]
18. Gonçalves, E.P.R.; Boaventura, R.A.R.; Mouvet, C. Sediments and aquatic mosses as pollution indicators for heavy metals in the Ave river basin (Portugal). *Sci. Total Environ.* **1992**, *114*, 7–24. [CrossRef]
19. Fernández, J.A.; Carballeira, A. Evaluation of contamination, by different elements, in terrestrial mosses. *Arch. Environ. Contam. Toxicol.* **2001**, *40*, 461–468. [CrossRef]

20. Okedeyi, O.O.; Dube, S.; Awofolu, O.R.; Nindi, M.M. Assessing the enrichment of heavy metals in surface soil and plant (Digitaria eriantha) around coal-fired power plants in South Africa. *Environ. Sci. Pollut. Res.* **2014**, *21*, 4686–4696. [CrossRef]
21. Müller, G. Index of geoaccumulation in sediments of the Rhine River. *Geol. J.* **1969**, *2*, 108–118.
22. Jorfi, S.; Maleki, R.; Jaafarzadeh, N.; Ahmadi, M. Pollution load index for heavy metals in Mian-Ab plain soil, Khuzestan, Iran. *Data Br.* **2017**, *15*, 584–590. [CrossRef] [PubMed]
23. Williams, C.R.; Harrison, R.M. Cadmium in the atmosphere. *Experientia* **1984**, *40*, 29–36. [CrossRef]
24. Adamiec, E.; Jarosz-Krzemińska, E.; Wieszała, R. Heavy metals from non-exhaust vehicle emissions in urban and motorway road dusts. *Environ. Monit. Assess.* **2016**, *188*, 369. [CrossRef] [PubMed]
25. Hjortenkrans, D.S.T.; Bergbäck, B.G.; Häggerud, A.V. Metal emissions from brake linings and tires: Case studies of Stockholm, Sweden 1995/1998 and 2005. *Environ. Sci. Technol.* **2007**, *41*, 5224–5230. [CrossRef]
26. Denier van der Gon, H.A.C.; Hulskotte, J.H.J.; Visschedijk, A.J.H.; Schaap, M. A revised estimate of copper emissions from road transport in UNECE-Europe and its impact on predicted copper concentrations. *Atmos. Environ.* **2007**, *41*, 8697–8710. [CrossRef]
27. Yu, C.H.; Huang, L.; Shin, J.Y.; Artigas, F.; Fan, Z.H.T. Characterization of concentration, particle size distribution, and contributing factors to ambient hexavalent chromium in an area with multiple emission sources. *Atmos. Environ.* **2014**, *94*, 701–708. [CrossRef] [PubMed]
28. Sanderson, P.; Su, S.S.; Chang, I.T.H.; Delgado Saborit, J.M.; Kepaptsoglou, D.M.; Weber, R.J.M.; Harrison, R.M. Characterisation of iron-rich atmospheric submicrometre particles in the roadside environment. *Atmos. Environ.* **2016**, *140*, 167–175. [CrossRef]
29. Tian, H.Z.; Lu, L.; Cheng, K.; Hao, J.M.; Zhao, D.; Wang, Y.; Jia, W.X.; Qiu, P.P. Anthropogenic atmospheric nickel emissions and its distribution characteristics in China. *Sci. Total Environ.* **2012**, *417–418*, 148–157. [CrossRef]
30. Wong, G.W.K. Air pollution and health. *Lancet Respir. Med.* **2014**, *2*, 8–9. [CrossRef]
31. Barandovski, L.; Frontasyeva, M.V.; Stafilov, T.; Šajn, R.; Ostrovnaya, T.M. Multi-element atmospheric deposition in Macedonia studied by the moss biomonitoring technique. *Environ. Sci. Pollut. Res.* **2015**, *22*, 16077–16097. [CrossRef]
32. Rutkowski, P.; Diatta, J.; Konatowska, M.; Andrzejewska, A.; Tyburski, Ł.; Przybylski, P. Geochemical referencing of natural forest contamination in Poland. *Forests* **2020**, *11*, 157. [CrossRef]
33. Salo, H.; Bućko, M.S.; Vaahtovuo, E.; Limo, J.; Mäkinen, J.; Pesonen, L.J. Biomonitoring of air pollution in SW Finland by magnetic and chemical measurements of moss bags and lichens. *J. Geochem. Explor.* **2012**, *115*, 69–81. [CrossRef]

Publisher's Note: MDPI stays neutral with regard to jurisdictional claims in published maps and institutional affiliations.

Review

Atmospheric Deposition and Element Accumulation in Moss Sampled across Germany 1990–2015: Trends and Relevance for Ecological Integrity and Human Health

Angela Schlutow [1], Winfried Schröder [2],* and Stefan Nickel [3]

[1] ÖKO-DATA— Ecosystem Analysis and Environmental Data Management, Lessingstraße 16, 16356 Ahrensfelde, Germany; angela.schlutow@oekodata.com

[2] Landscape Ecology, University of Vechta, P.O.B. 1553, 49364 Vechta, Germany

[3] PlanWerk, Unterdorfstraße 3, 63667 Nidda, Germany; s.nickel@planwerk-nidda.de

* Correspondence: winfried.schroeder@uni-vechta.de

Abstract: Deposition of N and heavy metals can impact ecological and human health. This state-of-the-art review addresses spatial and temporal trends of atmospheric deposition as monitored by element accumulation in moss and compares heavy metals Critical Loads for protecting human health and ecosystem's integrity with modelled deposition. The element accumulation due to deposition was measured at up to 1026 sites collected across Germany 1990–2015. The deposition data were derived from chemical transport modelling and evaluated with regard to Critical Loads published in relevant legal regulations. The moss data indicate declining nitrogen and HM deposition. Ecosystem and human health Critical Loads for As, Ni, Zn, and Cr were not exceeded in Germany 2009–2011. Respective Critical Loads were exceeded by Hg and Pb inputs, especially in the low rainfall regions with forest coverage. The Critical Load for Cu was exceeded by atmospheric deposition in 2010 in two regions. Human health Critical Loads for Cd were not exceeded by atmospheric deposition in 2010. However, the maximum deposition in 2010 exceeded the lowest human health Critical Load. This impact assessment was based only on deposition but not on inputs from other sources such as fertilizers. Therefore, the assessment should be expanded with regard to other HM sources and specified for different ecosystem types.

Keywords: bioaccumulation; biomonitoring; critical loads; deposition forests; chemical transport modelling; geographic information system; heavy metals; mapping

Citation: Schlutow, A.; Schröder, W.; Nickel, S. Atmospheric Deposition and Element Accumulation in Moss Sampled across Germany 1990–2015: Trends and Relevance for Ecological Integrity and Human Health. *Atmosphere* **2021**, *12*, 193. https://doi.org/10.3390/atmos12020193

Academic Editor: Daniele Contini

Received: 14 January 2021

Accepted: 26 January 2021

Published: 31 January 2021

1. Introduction

Emissions of elements from natural and anthropogenic sources come down to earth as wet, occult [1,2], or dry deposition at locations distant from their origin where they accumulate in biomass and soils [3]. The geographical pattern of element deposition and accumulation is influenced by chemical and physical element characteristics, meteorological and topographical conditions, land use, and vegetation structure. Potential impacts on human health and ecosystems integrity through heavy metal (HM) accumulation in food chains, and acidification and eutrophication of soils and limnic ecosystems [4–6] are intended to be avoided through the Convention on Long-Range Transboundary Air Pollution [7,8] addressing Cd, Pb, and Hg, as well as N and S. The European Monitoring and Evaluation Programme (EMEP) is to collate emission data, to collect atmospheric deposition Europe-wide by technical devices, and to calculate and map atmospheric deposition by chemical transport models such as the Long Term Ozone Simulation—EURopean Operational Smog model (LOTOS-EUROS) and EMEP [9–21]). This monitoring and modelling data can be validated and complemented by monitoring the bioaccumulation of elements in moss [22–24]. In Europe, HM (since 1990), N (since 2005), and persistent organic pollutants (POP; since 2010) were determined in moss specimens sampled in a rather dense spatial pattern. Since 2000, this European Moss Survey (EMS) is part of the International Cooperative Programme on Effects of Air Pollution on Natural Vegetation and Crops (ICP Vegetation). Since 1990, the EMS was conducted every five years and covered up to 7300 sampling sites in up to 36 European countries enabling to map spatial patterns of bioaccumulation and to derive deposition estimates by regression modelling [25–29].

High concentrations of atmospheric pollutants can result in exceeding Critical Levels of atmospheric concentrations and Critical Loads (CL) of atmospheric deposition. CL are defined as quantitative estimates of exposure to one or more pollutants deposited from air to the ground below which significant harmful effects on specified sensitive elements of the environment do not occur according to present knowledge. Critical levels are defined as concentrations of pollutants in the atmosphere above which direct adverse effects on receptors, such as human beings, plants, ecosystems, or materials, may occur according to present knowledge [7,8].

The chemical elements regarded in this investigation are given in Figure 1. The data on element concentrations in moss collected in Germany were analysed by a broad range of statistical methods focusing among others on following five key issues:

1. Bivariate and multiple correlations with ecological features of the sampling sites and of their environment—for instance atmospheric deposition, canopy drip effects, and land use—with results from other biomonitoring programmes and with results from deposition measurements using technical collectors and deposition modelling;
2. Geostatistical analysis and surface-covering estimation and mapping of site-specific data;
3. Computation and geostatistical mapping of percentile statistics of element-, site-, and survey-specific measurements;
4. Calculation and geostatistical mapping of elements integrating indices and surveys;
5. Assessing the relevance of modelled HM deposition for ecosystem integrity and human health based on CL.

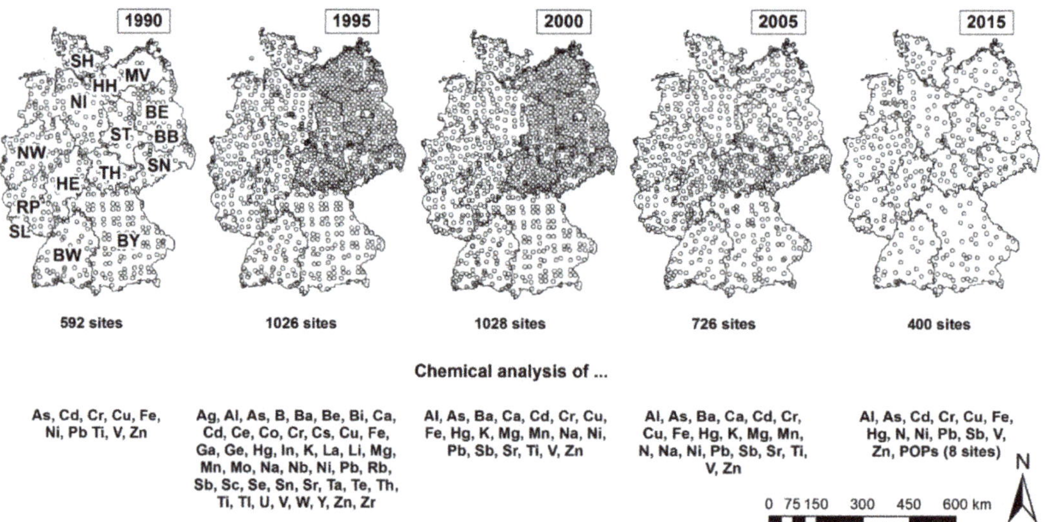

Figure 1. Sampling sites and elements regarding the German Moss Surveys 1990, 1995, 2000, 2005, and 2015. SH = Schleswig-Holstein; MV = Mecklenburg-West Pomerania; HH = Hamburg; NI = Lower Saxony; BE = Berlin; ST = Saxony-Anhalt; BB = Brandenburg; NW = North Rhine-Westphalia; SN = Saxony; TH = Thuringia; HE = Hesse; RP = Rhineland Palatinate; SL = Saarland; BY = Bavaria; BW = Baden-Wuerttemberg.

This article aims at presenting the current state of knowledge on atmospheric HM deposition and bioaccumulation and the assessment of its relevance for ecosystem integrity and human health in Germany. Both these aspects of the relevance of HM input are inter-linked. For example, excessive pollution of arable and grassland ecosystems contributes to the exposure on humans via the food chain. Damage to forest ecosystems reduces their recreational effect on humans. The article concentrates on two of the five key issues investigated in the framework of the German moss surveys: 1. Summarising EMS data collected from 1990 to 2015 across Germany by percentile statistics and calculation of elements and surveys integrating index scores 2. Reporting on the latest assessment of atmospheric heavy metal deposition with regard to ecological integrity and human health in Germany.

2. Materials and Methods

2.1. Bioaccumulation of Atmospheric Deposition of HM in Moss

Sampling and chemical analysis of moss specimens as well as classification and mapping of element concentrations determined follow a harmonised methodology (for EMS 2015 refer to ICP Vegetation [27]). Between 1990 and 2015, the number of moss sampling sites in Europe ranged between 4499 and 7312 in 20–36 countries. In the German Moss Survey, moss specimens were collected at 592 (1990 [30]), 1026 (1995 [31,32], 1028 (2000 [28]), 726 (2005 [33]), and 400 (2015 [34]) sites in forested areas. The reduction of sampling sites was performed according a statistically sound methodology [33,35]. Germany did not take part in the EMS 2010.

The international classification of element concentrations in moss [27] is too coarse to display the spatial variance of decreasing element concentrations. Additionally, the extensive data on up to 40 metal elements collected between 1990 and 2015 every five years at up to 1028 sites across Germany were summarised as far as possible in terms of a multi-metal index (MMI). Thereby, mapping of spatial and temporal trends was preserved. To this end, the element-specific data on HM accumulation was divided into 10 percentiles, which were then transformed into MMI score values ranging from 1 to 10.

The statistical analyses presented in this review regard HM concentrations that were measured in moss specimens collected in Germany 1990–2015. The following percentile statistics were calculated and mapped for spatial point data as well as for geostatistical surface estimations [23]:

1. Element- and survey-specific quantiles (10 classes as defined by the 10th, 20th,100th percentile for each of the surveys) enabling to detect whether the geographical patterns of bio-accumulation hot spots of 7 elements (Cr, Cu, Fe, Ni, Pb, V, Zn: 1990–2015) and 12 elements (Al, As, Cd, Cr, Cu, Fe, Hg, Ni, Pb, Sb, V, Zn: 1995–2015) from previous campaigns remain hot spots even with decreasing atmospheric deposition and bio-accumulation, or whether the bio-accumulation patterns shift across time or not.
2. Element-specific quantiles integrating all surveys 1990–2015 allowing for a statistically derived differentiation of 7 (1990–2015) and 12 (1995–2015), respectively, element concentrations into 10 classes and for their mapping across time in time and space despite decreasing element concentrations.
3. Seven and 12 elements and surveys, respectively, integrating calculation of a Multi Metals Index (MMI90-2015: 7 HM; MMI95-2015: 12 HM). To this end, gridded data on element concentration in moss were each subdivided into 10 percentile classes (0–10th percentile, > 10th to 20th percentile, ... > 90th to 100th percentile). In a second step, scores are assigned to the element-specific percentile classes (0–10th percentile = index value 1, > 10th to 20th percentile = index value 2, and so on). To calculate the MMI ranging from 1 (low metal accumulation) to 10 (high metal accumulation), the element-specific index values for each object were averaged.

The results for Element-specific quantiles integrating all surveys 1990–2015 (b) and MMI90-2015 and MMI95-2015 scores are presented in Section 3.1.

2.2. Assessing Impacts of Atmospheric Deposition

2.2.1. Assessment Values

The second aim of this contribution was to assess potential HM deposition effects on ecosystems and human health on the basis of legal requirements and environmental quality objectives. As HM can be transported through the atmosphere over long distances and across national borders, both national and international regulations and assessment methods were considered thereby. The regulations and recommendations compiled in Tables 1 and 2 contain different categories of assessment values, which differ with regard to their protective purpose, the respective level of protection, and protective objective. For this reason, this study uses the overarching term "assessment value" but takes over the nomenclature of quotations from the rules and regulations. In addition, a distinction is made between precautionary assessment values and those which serve to avert danger. Precautionary assessment values indicate limits of resilience (concentrations in environmental compartments or substance flows) below which there is no concern of significant impairment of ecosystems and their functions and services to humans. They apply generally, i.e., beyond the sphere of influence of concrete facilities, projects, or management measures, and they are independent of usage claims. In law, the concept of danger is always linked to a certain probability of the occurrence of significant, harmful changes. In principle, assessment values that serve to avert hazards permit higher pollutant concentrations or inputs than precautionary ones. As a rule, they serve to assess concrete (including planned) facilities, projects, or management measures and are derived from specific uses (e.g., test values and measure values in soil protection). Table 1 compiles the assessment values used in this study to compare them with CL. Due to the methodological differences in their derivation, they are only comparable to each other to a limited extent and with CL. The differences, some of which are clear, are due to different levels of protection, protection objectives, and the relationship between effects (Table 2).

Table 1. Assessment values for heavy metal fluxes (g ha^{-1} a^{-1}) for the protection of ecosystems and human health.

Metal	TA Luft [1]	TA Luft [2]	BBodSchV [3]	39th BImSchV [4,5]	Directive 2004/107/EC [5]	Directive 2008/50/EC [5]
		Emitter-Related			General Load	
Hg	4	110 [F], 11 [G]	1.5			
Cd	7	9 [F], 117 [G]	6	4.4 [H], 7 [C], 4 [D], 2.5 [FG]	4.4 [H], 7 [C], 4 [D], 2.5 [FG]	
Pb	365	675 [F], 6935 [G]	400	435 [H], 716 [C], 420 [D], 250 [FG]		435 [H], 716 [C], 420 [D], 250 [FG]
As	15	4271 [F], 219 [G]		5.2 [H], 6 [C], 4 [D], 2.2 [FG]	5.2 [H], 6 [C], 4 [D], 2.2 [FG]	
Ni	55		100	17.4 [H], 28 [C], 17 [D], 10 [FG]	17.4 [H], 28 [C], 17 [D], 10 [FG]	
Cu			360			
Zn			1200			
Cr			300			
Tl	7	26				

[1] TA Luft [36] = Technical Instructions for Air pollution control (deposition values to protect human health). [2] TA Luft [35] = Technical Instructions for Air pollution control (deposition values as reference points for the special case examination to protect environment). [3] BBodSchV [37] = Federal Soil Protection Ordinance (permissible additional load according to §11 para. 2). [4] 39th BImSchV [38] = 39th Federal Immission Control Ordinance. [5] Converted from assessment values for concentrations (Tables 33 and Table 34 in [17]) published in Directive 2004/107/EC [39] and Directive 2008/50/EC [40]. [F] For field. [G] For grassland. [C] For coniferous forest. [D] For deciduous forest. [H] For housing settlement.

Table 2. Compilation of categories of assessment values from legal, sublegal regulations, and recommendations for air pollution control for heavy metal fluxes for the protection of ecosystems, protected goods, levels and objectives, and impact indicators.

Sources for Appraisal Values	Designation/Category of Appraisal Values	Binding Force	Objects of Protection	Level of Protection	Application for the Assessment	Impact Indicator
39th BImSchV [38]	Immission limit value (Pb)	Legally binding	Man + Environment	Precaution + danger prevention	General strain	Human toxicological impact thresholds
	Target values (As, Cd, Ni)	Not legally binding	Man + Environment	Precaution + danger prevention	General strain	Human toxicological impact thresholds
TA Luft [36]	Immission values for pollutant deposition	Binding on administra-tive action	Environment	Hazard prevention (immission values)	Plants requiring approval	Human toxicological impact thresholds
Directive 2004/107/EC [39]	Immission target values	EU recommendation	Man and environment (via soil, plants)	Precaution + danger prevention	General strain	Human toxicological impact thresholds
Directive 2008/50/EC [40]	Immission limit value (Pb)	Legally binding	Man + Environment	Precaution + danger prevention	General strain	Human toxicological impact thresholds
CLRTAP [7,8]	Critical Loads (CL(M)$_{eco}$) (Section 2.2.2)	Recommendation/ orientation	Terrestrial ecosystems, soil organisms and plants	Precaution	General strain	Ecotoxicological thresholds NOEC, LOEC microorganisms, invertebrates and plants
	Critical Loads (CL(M)$_{drink}$), (CL(Cd)$_{food}$) (Section 2.2.2)	Recom-mendation/ orientation	Human	Precaution	General strain	Limit values of Drinking Water Ordinance and critical limit for Cd in wheat
BBodSchV [37]	Precaution-ary values	Legally binding	Ecosystems, soil organisms and plants	Precaution	General validity (cross-use)	Ecotoxicological thresholds NOEC, LOEC, (in future: HC5, EC10) of soil organisms and plants (all pathways) + Background values
	Permissible annual additional load	Legally binding *	Ecosystems, soil organisms and plants (all paths of action)	Pension entitlement and limited in the long run	General validity (cross-use)	Information on the amount of ubiquitous deposition

* It is legally binding that values for the permissible additional load are derived. The values themselves are rather indicative, as there are no concrete prescribed applications. Installations or input values due to management are subject to other technical laws. BBodSchV [38] = Federal Soil Protection Ordinance; CLRTAP [7,8] = Convention on Long-range Transboundary Air Pollution; EC10 = ECx is the effect concentration at which x% effect (mortality, inhibition of growth, reproduction, …) is observed compared to the control group; HC5 = Hazardous concentration for 5% of the species; LOEC = Lowest Observed Effect Level; NOEC = No Observed Effect Concentration; PNEC = Predicted No Effect Concentration; TA Luft [36] = Technical Instructions Air.

The protection of human health and ecosystems and their functions against adverse effects from air pollutant deposition is generally ensured if HM inputs are completely avoided. However, this is currently not a realistic assumption. On the basis of empirical evidence, it is assumed that the protection of these objects may be reached if specific critical concentrations or loads of HM in environmental media are not exceeded. Thereby, only with the calculation of CL the balance between inputs and outputs can be proved.

2.2.2. Basics for the Determination of Critical Loads for Heavy Metal Deposition

CL for Cd, Pb, and Hg have already been calculated for the entire EMEP region. They serve as policy advice, in particular to examine and justify whether further emission reductions are necessary. To date, they have not been designed as binding air concentration or deposition values. CL indicate the total input rate below which adverse effects on ecosystems and human health (paths atmosphere-soil-groundwater for drinking water use and atmosphere-soil-food wheat (only for Cd)) can be excluded in the long term according to current knowledge. Consequently, if CLs are complied with, risk minimisation is achieved below the classic danger threshold, which means that the assessment values are very precautionary.

The CL concept focuses on the budgets of substances in ecosystems. Ecosystem specific features (soil, climate, use, etc.) are taken into account when calculating the critical load values. As a result, there is not only one CL, but rather a range of values that allows a comprehensive, regionalised representation of the sensitivity of ecosystems, food crops, and drinking water to HM.

In addition to natural and semi-natural ecosystems, agricultural land is also considered both as ecosystems and as areas where human and eco-toxicological values must be respected. CL aimed at protecting ecosystems are hereinafter referred to as CL(M)eco. CL aiming at protecting human health, e.g., drinking water, are abbreviated CL(M)drink and those aimed at protecting food for humans CL(M)food, where (M) stands for heavy metal and can be replaced by the respective element symbol (Cd, Pb, Hg, . . .). The determination of CL(M)eco was based exclusively on eco-toxicological threshold values. This means that the CL(M)eco were determined on the basis of effects. Experimentally determined zero effect threshold values (NOEC or PNEC) were used as "critical limits" in the calculation of the CL(M)eco. For the CL(M)drink, internationally agreed critical concentrations were used in drinking water and for CL(Cd)food in food wheat.

For Germany, an assessment of the input rates into ecosystems in the equilibrium of inputs and outputs will be carried out according to the CL concept [34]. Their mapping for Germany is carried out on a scale of 1:1 million and provides an overview of the sensitivity of terrestrial ecosystems to nine HM. Ecosystem integrity and human health are regarded as protection goals.

By definition, CL for HM are the highest total input rate of HM under consideration (from atmospheric deposition, fertilisers, and other anthropogenic sources) below which no long-term adverse effects on human health and on the structure and function of ecosystems are to be expected according to the current state of knowledge [7,8]. CL are calculated according to the mass balance approach assuming a chemical equilibrium in the system under consideration and a steady state at a concentration level defined by the critical limit. This is an impact-based derived limit concentration in certain ecosystem compartments below which significant adverse effects on human health as well as on defined sensitive components of ecosystems can be excluded according to the current state of knowledge.

Cd has been identified as an important pollutant in relation to the maintenance of food quality for the protection of human health. With this metal, uptake from the soil into the vegetation is comparatively high, so that accumulations in the soil entail the potential danger of health effects via plant food. Wheat was selected as the indicator plant. Wheat grain accounts for a significant proportion of food in Germany (as in Europe) and its cultivation accounts for a large proportion of agricultural land in Germany (and other

European countries) [41]. CLs for the protection of drinking water are mapped for all ecosystem types. CL were therefore determined for three objects of protection:

- CL(M)eco: Critical Load for a metal (M stands for As, Cd, Cu, Cr, Hg, Ni, Pb, Zn,) to protect the sensitive biota of the ecosystem;
- CL(M)drink: Critical Load for a metal (M stands for As, Cd, Cu, Cr, Hg, Ni, Pb, Zn) for Protection of drinking water for human beings;
- CL(Cd)food: Critical Load for Cd for the protection of arable crops (here: wheat-producing as a food for human beings.

2.2.3. Calculation of Critical Loads for Heavy Metals in Germany

The methodological approach for the calculation of CL for HM in this study follows [7,8] (Chapter V.5). All relevant fluxes into or from a certain soil layer, in which the essential substance conversions occur or in which the receptors have their distribution focus and which is therefore relevant for the effects in the system, were compared. The consideration of HM fluxes, reserves, and concentrations refers to mobile or potentially mobilizable metals, only they are relevant for the consideration of substance fluxes.

The mass balance equation includes as output paths from the terrestrial ecosystem the uptake into the biomass with subsequent harvest and the output with the leachate flow as follows:

$$CL(M) = M_u + M_{le(crit)}$$

where:

$CL(M)$ = Critical Load of the metal M (g ha^{-1} a^{-1})

M_u = Net uptake of the metal M into harvestable plant parts (g ha^{-1} a^{-1}).

$M_{le(crit)}$ = Tolerable (critical) leaching of the metal M from the considered soil layer with exclusive consideration of vertical rivers (leachate) (g ha^{-1} a^{-1}).

The inclusion of further terms was in accordance with the recommendations of the Expert Panel for HM to the ICP Modelling & Mapping [41,42]. For the CL calculation, the necessary data were spatially linked with GIS software ArcView, 10.2.1 and transferred to an Access 2000 database. Both original data such as precipitation and derived data such as values for the organic matter content (OM) and pH values derived from the land use-differentiated soil overview map of Germany, 1:1,000,000 BÜK1000N were used. The storage, evaluation, and presentation of the data were performed in polygons, which result from the intersection of the input data.

Identifying the geographical distribution and magnitude of total (the sum of wet and dry) deposition of atmospheric pollutants is crucial to determine the areas, populations, ecosystems, and farmlands that are most vulnerable to its negative effects and that would most benefit from measures to control excessive pollutant loads. To this end, the relevance of HM deposition for ecological integrity and human health in terms of CL were compared to the respective atmospheric deposition modelled with the chemical transport model LOTOS-EUROS. CL and CL exceedances by modelled atmospheric deposition were mapped for Germany on a scale of 1:1 Mio.

The deposition dataset [18] contains information on the concentrations and deposition fluxes of various HM (As, Cd, Cr, Cu, Ni, Pb, V, Zn) for the years 2009, 2010, and 2011. The deposition data sets (dry and wet) were combined to the total deposition, converted into the unit of measure of the critical load [g ha^{-1} a^{-1}] and blended with the critical load receptor areas. The CL exceedance rates were calculated as follows:

$$MinExcCL(M)_{eco} = (MinM_{dep} + MinM_{fertilizer}) - CL(M)_{eco}$$

and

$$MaxExcCL(M)_{eco} = (MaxM_{dep} + MaxM_{fertilizer}) - CL(M)_{eco}$$

where:

$MinExcCL(M)_{eco}$ = Minimum ecosystem critical loads exceedance in the German receptor areas due to total deposition from the air and fertiliser inputs

$MaxExcCL(M)_{eco}$ = Maximum ecosystem critical loads exceedance in the German receptor areas due to total deposition from the air and fertiliser inputs

$MinM_{dep}$ = Minimum of the total deposition from the air in the German receptor surfaces, corresponds to the highest minimum of the three years 2009–2011

$MaxM_{dep}$ = Maximum of total deposition from the air in the German receptor surfaces corresponds to the highest maximum of the three years 2009–2011

$MinM_{fertilizer}$ = Minimum of the metal inputs with the fertilization in the German receptor surfaces

$MaxM_{fertilizer}$ = Maximum of the metal inputs with the fertilization in the German receptor areas

$CL(M)_{eco}$ = Median of the Ecosystem Critical Loads for the metal in the German receptor surfaces

2.2.4. Modelling Heavy Metal Deposition

The deposition dataset was produced by use of the chemical transport model LOTOS EUROS [18] and contains information on the atmospheric concentrations and deposition of various HM (Cd, Pb, Ni, As, Zn, Cu, V, and Cr) for the years 2009, 2010, and 2011. The data include information on the centroid coordinates of the degree cells with the deposition data. The modelled data for Cd and Pb were provided in the $0.125° \times 0.0625°$ network of the Long/Lat system and the data for the other HM in the $0.5° \times 0.25°$ resolution. The methodology of the deposition calculation is explained by Schaap et al. [18], where the different scales of the deposition data were also justified. The available point data (centroids of the grid cells and intersection points of the grid cells) were assigned to the planar degree grid, which has the corresponding mesh size of the respective resolution. This step was necessary to compare the deposition data with the critical load data set.

The datasets on dry and wet atmospheric deposition were combined to the total deposition, converted into the unit of measure of the CL ($g\ ha^{-1}\ a^{-1}$) and blended with the critical load receptor areas covering Germany. After a detailed examination of all years (2009, 2010, and 2011) for all heavy metals, the data for 2010 turned out to be the highest. For this reason, the full-year presentation is limited to 2010.

3. Results and Discussion

3.1. Trends of HM Bioaccumulation Integrating Metal Elements and Surveys 1990–2015

The surveys integrating metal concentrations classified by use of percentile statistics (Section 2.1.) were transformed to scores ranging from 1 to 10 and aggregated for each of the 3 km by 3 km grid cells covering Germany. Based on this procedure a Multi Metal Index (MMI) encompassing all data collected in the framework of EMS 1990 to 2015 and integrating Cr, Cu, Fe, Ni, Pb, V, and Zn (MMI90-2015) and MMI95-2015 (Al, As, Cd, Cr, Cu, Fe, Hg, Ni, Pb, Sb, V, Zn) was calculated and mapped (Figure 2).

In the following, the spatial patterns depicted in Figure 2 are summarized and the median MMI values for Germany are referred to. Thereby, the geostatistically estimated median MMI are added in squared brackets by those calculated from the sample point measurements. In 1990, almost the whole territory of Germany is covered by MMI90-2015 values exceeding MMI = 6.0 (median 8.3 [7.7]). The survey 1995 (median = 7.3 [6.7]) indicated a slight area-wide decline of the MMI compared with that conducted in 1990. This trend was continued in the survey 2000 (median = 4.6 [4.4]). However, from 2000 to 2005, this trend changed and the MMI increased (median = 5.3 [5.1]) due to increasing bioaccumulation of Cr, Hg, Sb, and Zn. Until the survey in 2015, again a Germany-wide decrease of MMI90-2015 could be corroborated (median = 1.7 [2.0]). In 2015, areas with MMI90-2015 exceeding 4.0 could only be determined for North Rhine-Westphalia and the upper Rhine valley (Baden-Wuerttemberg) (Figure 2 Above).

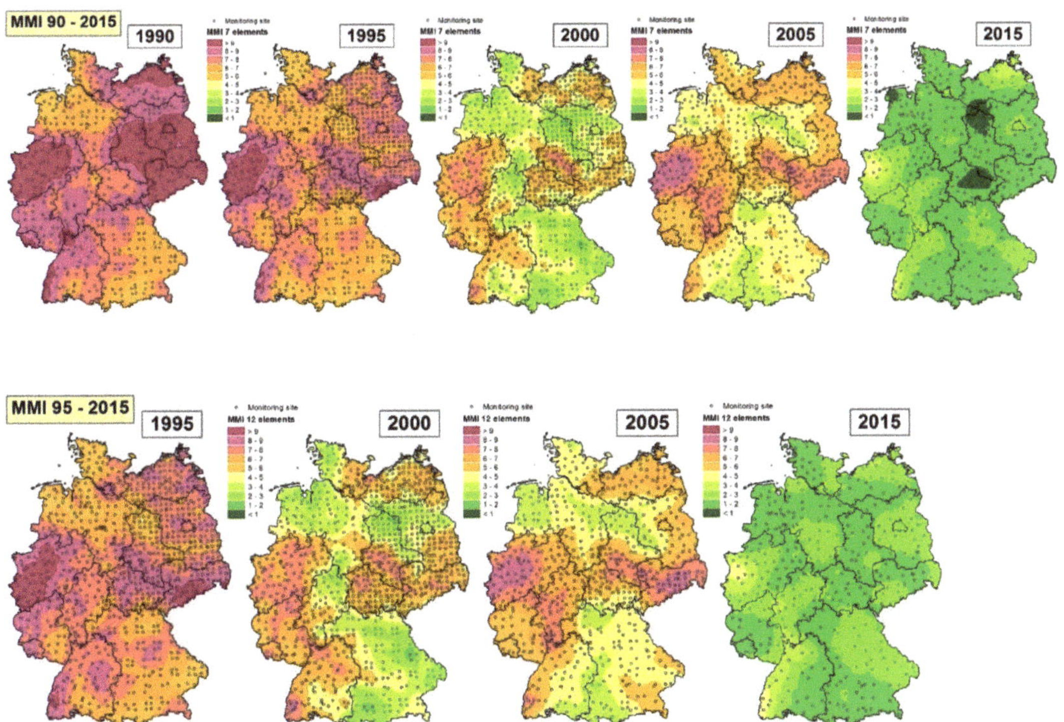

Figure 2. Spatial patterns of Multi Metal Index (MMI) integrating Cr, Cu, Fe, Ni, Pb, V, and Zn based on data collected from 1990 to 2015 (**Above**). Spatial patterns of MMI integrating Al, As, Cd, Cr, Cu, Fe, Hg, Ni, Pb, Sb, V, and Zn based on data collected from 1995 to 2015 (**Below**).

The maps depicting the spatial patterns MMI95-2015 integrating Al, As, Cd, Cr, Cu, Fe, Hg, Ni, Pb, Sb, V, and Zn measured in moss sampled 1995–2015 (Figure 2 Below) also indicate a clear decrease between 1995 and 2000 from 7.5 [6.6] to 5, 1 [4.8]. From 2000 to 2005 the MMI90-2015 turned to 5.3 [5.0]. Finally, during the years 2005–2015 the MMI95-2015 decreased to 2.0 [2.3]. The spatial patterns and hot spots of MMI95-2015 are similar to MMI90-2015 with correlation coefficients (Spearman) between r_s = 0.83 (year 2015, $p < 0.01$) and r_s = 0.98 (year 2000, $p < 0.01$) and highest values in the Ruhr region (North Rhine-Westphalia) and in the upper Rhine valley (Baden-Wuerttemberg).

The element and surveys integrating trends of bioaccumulation of HM atmospheric deposition depicted in Figure 2 and described the two preceding paragraphs can be explained by looking at the element- and survey-specific quantiles (Section 2.1.) and at the element-specific quantiles integrating surveys (Section 2.1.). From that we can derive information identifying the elements with a decreasing but discontinuous trend (Table 3).

N, which is not in the focus in this review, did not show any statistically significant time trend from 2005 to 2015 but a change of the geographical location of its hot spots. Regarding the development of Cd, Cr, Hg, and Pb accumulation in moss since 2005, a Germany-wide decrease could be detected and proved as statistically significant ($p < 0.05$). In terms of median values, the decline of HM bioaccumulation ranged between −4% (Hg) and −50.4% (Pb). The same tendency could be corroborated for comparisons between EMS 2015 values with those in the first year of sampling. Since 1990 (in case of Hg since 1995) the median values of all HM were reduced significantly. The most distinct decline was found for Pb (−85.9%) and the lowest for Hg (−20%). The trends for Cd and Pb are in line with respective emission data covering the period 1990 to 2015 [43]. For Cd, the Spearman correlation coefficient rs > 0.3

($p < 0.01$) was also found between concentrations in moss sampled in 2005 and according modelled atmospheric deposition, which was calculated by use of LOTOS-EUROS chemical transport modelling. The same holds true for Pb (Section 3.2). According to [43] emissions from metallurgy (Cd, Ni, and Pb), power economy (As, Cd, Ni, and Pb), manufacturing and constructing industry (As, Ni, and Pb), and traffic (Pb, due to the compulsory introduction of unleaded petrol in 1988 in Germany and 2000 in the EU) declined since 1990. However, the decrease of Hg bioaccumulation is less than the reduction of Hg emissions [43]. This is possibly due to long-range transport of gaseous Hg and atmospheric residence times of 6 to 18 months [18]. The concentrations of Cu and Zn in moss contradict the emission trends. At least for Cu, the Spearman correlation between the concentration in moss collected in 2005 and the modelled atmospheric deposition is rs 0.22 ($p < 0.01$), for Zn, respectively, rs < 0.2 ($p < 0.01$). For Cr, good agreement could be identified between the emission trends [43] and the concentrations in moss during the period 1990 to 2015. Strikingly, the Cr concentrations in moss were extraordinarily high in 2005, especially in Mecklenburg-West Pomerania and conurbations such as Bremen, Hamburg, Dresden, Halle/Leipzig, and the Ruhr region. Respective increased values were also reported from Austria.

Table 3. Element-specific trends of heavy metal (HM) bioaccumulation in moss (Germany, 1990–2015).

Heavy Metal	Survey(s)	Trend
Pb, Fe	1990–2015	Continuous reduction of concentrations
	2015	Only areas with very low concentrations *
Cr, *Sb*, Zn	1990 *(1995)*–2015	Statistically significant decrease
	2000–2005	Interim increase
	2015	Only areas with very low concentrations *
Al, As, Cd, Cu, *Hg*, Ni, V	1990 *(1995)*–2015	Statistically significant decrease
	2000–2005	Intermediate standstill

Al, As, Cd, Hg, Sb were monitored since *1995*, Cr, Cu, Fe, Ni, Pb, V, and Zn since 1990. * The international classification of element concentrations in moss [27] does not allow displaying spatial variance with decreasing element concentrations. Therefore, the data on HM accumulation was divided into ten percentile classes (Section 2.1).

3.2. Atmospheric Heavy Metal Deposition

The ranges of modelled HM deposition in Germany for the year 2010 [18] are shown in Table 4. For Hg, deposition data in Germany are only available as modelled EMEP data 2013 on a 50×50 km^2 grid (EMEP 2015) 5–95 [10,11]. The ranges of the 5th percentile to 95th percentile vary from 0 to 0.76 g ha^{-1} a^{-1} for deciduous forest, from 0.16 to 0.87 g ha^{-1} a^{-1} for coniferous forest, from 0.8 to 0.35 g ha^{-1} a^{-1} for acre, from 0 to 0.31 g ha^{-1} a^{-1} for marshes, from 0.08 to 0.34 g ha^{-1} a^{-1} for grassland, and from 0.08 to 0.34 g ha^{-1} a^{-1} for grassland. For Thallium deposition data are not available for the whole of Germany, but only from a few measuring stations.

3.3. Heavy Metal Inputs from Other Sources

In addition to the atmospheric inputs of HM presented in Section 3.2, the following other input pathways play an important role in soil pollution. The respective values were collected by Knappe et al. [44] from nationwide surveys and they are given in Table 5:

- Application of mineral and organic ("farm") fertilisers containing HM on agricultural land (arable land and intensive grassland);
- Application of pesticides containing HM on agricultural land;
- Application of lime fertilizers containing HM in forests.

Table 4. Background atmospheric total deposition of heavy metals (g ha^{-1} a^{-1}) [18].

Statistical Parameter	Pb	Cd	As	Ni	Cu	Zn	Cr	V
5. Perc.	4.96	0.21	0.282	1.98	3.07	11.89	0.84	0.33
25. Perc.	5.90	0.26	0.333	2.35	4.81	16.07	1.03	0.39
50. Perc.	6.71	0.29	0.380	2.69	5.89	19.08	1.22	0.44
75. Perc.	7.81	0.33	0.437	3.10	7.17	22.24	1.45	0.52
95. Perc.	11.00	0.45	0.603	3.92	10.67	33.38	2.08	0.85
Min.	3.59	0.17	0.208	1.42	1.98	8.24	0.66	0.30
Max.	87.25	2.33	1.026	7.11	29.42	76.63	3.97	1.90
Mean	7.24	0.31	0.401	2.80	6.42	20.25	1.31	0.49

Table 5. HM inputs from fertilisation in agriculture and forestry (in g ha^{-1} a^{-1}) in Germany [44].

Heavy Metal		Acre				Grassland		Organic Agriculture	Forest
		Mineral Fertilizer	Compost	Sewage Sludge	Farm Fertilizer	Mineral Fertilizer	Farm Fertilizer	Farm Fertilizer	Liming
As	Min.	0.99	8.91	4.4	0.76	1.07	2.61	0.527	0.4
	Max	1.69	31.28	7.97	3.9	1.28	4.98	0.72	
Pb	Min.	4.19	82.82	47.98	1.81	8.11	8.65	0.859	0.7
	Max.	8.76	315.89	87.28	10.39	9.31	10.94	1.245	
Cd	Min.	1.33	1.61	1.14	0.4	2.05	0.66	0.619	0.2
	Max.	3.3	4.57	2.61	1.43	2.61	0.86	0.82	
Cr	Min.	46.61	73.42	55.45	40.65	24.42	21.51	8.82	8.2
	Max.	57.23	163.8	70.14	59.32	29.2	27.32	10.38	
Cu	Min.	11.27	98.61	282.21	8.8	20.27	81.43	0.78	0.8
	Max.	34.61	397.79	514.49	220.17	27.13	174.27	3.53	
Ni	Min.	6.64	31.83	26.76	5.15	5.4	8.32	1.75	1.6
	Max.	9.16	109.76	45.77	16.75	5.8	14.82	2.03	
Hg	Min.	0.01	0.27	0.64	0.01	0.03	0.06	0.037	0.06
	Max.	0.05	1.06	1.19	0.09	0.03	0.12	0.046	
Tl	Min.	0.08	0.26	0.33	0.04	0.11	0.13	0.08	0.09
	Max.	0.18	0.85	0.64	0.23	0.14	0.24	0.09	
Zn	Min.	66.76	376.05	694.12	31.09	99.35	331.6	9.73	4.2
	Max.	250.2	1445.75	1272.09	911.77	129.54	706.5	27.48	

3.4. Critical Load Exceedances Due to Atmospheric Deposition

After comparing the deposition data available from this study for 2009–2011 [18] with the respective CL, three exceedances occurred: Pb (drinking water and ecosystem protection) and Cu (ecosystem protection) (Figures 3–5). The critical load exceedance for Pb deposition from air with the protection target drinking water quality can only be expected on a relatively small receptor surface. Between 1.32% (2011) and 2.44% (2010) of the receptor surface have an increased risk for drinking water quality due to airborne pollutant deposition of Pb. Critical load exceedance with airborne Cu deposition with ecosystem integrity as a protection target is also relatively low. The values vary between 0.35% (2011) and 1.16% (2010) of the receptor areas. A significantly higher proportion of land is affected by a critical load exceedance in airborne Pb deposition with the protection objective of ecosystem integrity. Here, space shares between 5.18% (2011) and 14.36% (2010) are achieved. In absolute terms, this corresponds to 14,494 km^2 (2011) and 40,181 km^2 (2010). The focus is on the Leipzig and Thuringian basins, the loess areas (including the northern Harz foothills and the Lower Saxony area), the Erzgebirge foothills, the Lower Saxony highlands, parts of the Mecklenburg Lake District backcountry and parts of the Ruhr area (Lower Rhine lowlands and Cologne Bay). These areas are already identified as particularly sensitive to Pb deposition. Maps of the exceedance rates of the CL for Pb (drinking water and ecosystem protection) and Cu (ecosystem protection) are shown in Figures 3–5. There is no need to map exceedances of other CL as there are no exceedances of these CLs in 2010.

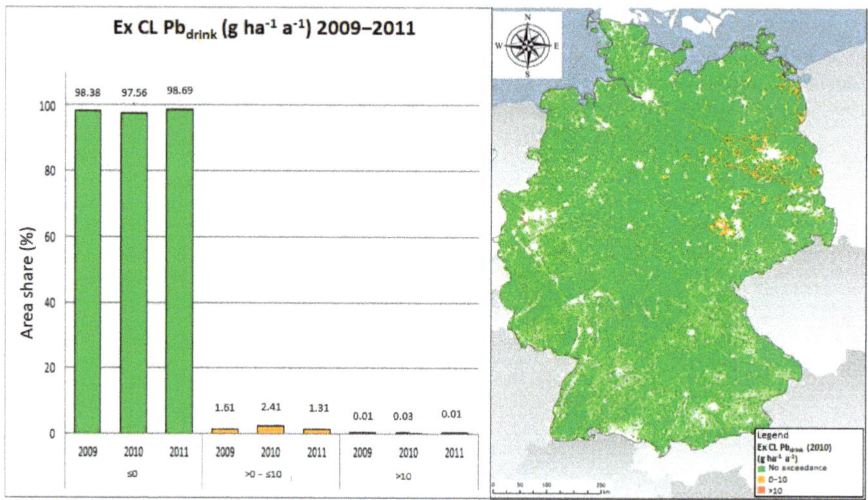

Figure 3. Exceedance of Critical Loads with the protection objective of drinking water quality by atmospheric Pb deposition.

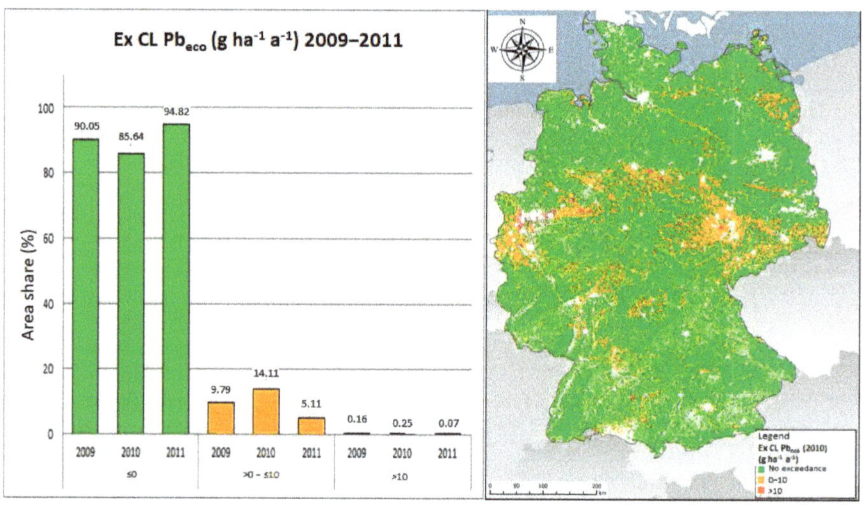

Figure 4. Exceedance of Critical Loads with the aim of protecting ecosystem integrity by atmospheric Pb deposition.

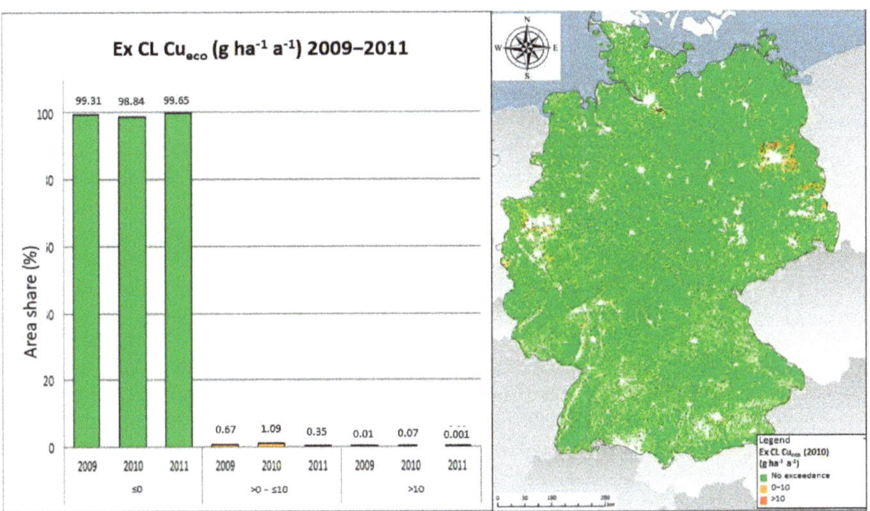

Figure 5. Exceedance of Critical Loads with the protection objective ecosystem integrity by atmospheric Cu deposition.

These results based on the deposition calculations with the LOTOS-EUROS model differ significantly from the deposition calculations with the EMEP model [18]. The LOTOS-EUROS model consistently results in lower medians and maxima than EMEP. The EMEP/LOTOS-EUROS ratios correlate very strongly with the deposition height, i.e., where EMEP calculates particularly high deposition (e.g., Pb in the Ruhr area or Cd in North Rhine-Westphalia), the differences between the two models are also highest. Conversely, the low deposition calculated with LOTOS-EUROS, e.g., at altitudes in southern Bavaria, is higher than with EMEP (Cd: 45% higher, Pb: 12% higher).

The integrative analysis of the LOTOS-EUROS models (2005, 2007–2011, Germany) with the geostatistical area estimates of the heavy metal contents in moss (EMS, 2005, Germany) showed stronger statistical correlations than the correlation with measured concentrations in moss [24]. For Cu, the correlation is weak. At Pb, the mean correlations to LOTOS-EUROS were stronger than to EMEP, but strongest to the arithmetic mean of LOTOS-EUROS and EMEP. It follows from this that if EMEP background deposition were applied, the proportion of areas with critical load exceedances would be higher. However, the comparison of the geostatistical area estimates of the heavy metal contents in moss with the EMEP results also leads to the conclusion that the LOTOS-EUROS results for the Pb deposition are closer to reality than the EMEP data and thus at least for Pb the exceedance rates shown above have lower uncertainties than when using EMEP deposition. The uncertainties of the LOTOS-EUROS results cannot be measured quantitatively, because there are not enough measurement data containing wet and dry deposition at the same time.

3.5. Statistical Evaluation of Critical Load Exceedances

In the following, the values for the total deposition for 2010 by Schaap et al. [18] and for Hg from the EMEP dataset for Germany 2013 are compared with the CL for 2016 (minimum, 5-percentile, 95-percentile and median). This comparison shows possible risks for ecosystems and human health for the areas that cannot be represented in the German data set due to its small scale. Since the 2010 deposition dataset shows higher average deposition than the 2009 and 2011 data sets for Germany, this comparison also tends to show more unfavourable conditions, so that the risk assessment is conservative.

The German datasets for CL and deposition were calculated on the basis of input data collected on a scale of 1:1 million. The mapping units of BÜK1000N, CORINE 2006 [45] and the site types classified by climate zones are not homogeneous in reality. They may contain sprays that are more sensitive to heavy metal ingress. The German dataset of CL 2016 [34] and the exceedances of CL in 2010 is therefore not applicable for large-scale area-based evaluations or even for site-specific statements. A rough orientation for individual sites can only be derived from the German dataset if it can be demonstrated that the same site conditions prevail for specific areas or sites as were used as a basis in the German CL(M) and deposition datasets. In order to nevertheless be able to derive a statement on the risks also for areas that cannot be represented at a scale of 1:1 million, the respective worst-case values are compared with each other, i.e., the lower range limit of the CL with the upper range limits of the deposition. It is assumed that the parameter values of the site types that are not representative in terms of area on a scale of 1:1 million and are not represented in the German data sets would not lie outside the ranges of the CL and atmospheric deposition in Germany.

The comparison of the ranges of CL and deposition shows the risk of whether and to what extent, in the worst case, further accumulation of HM in ecosystems or in groundwater or in the soil of wheat fields above the critical concentrations could take place if the annual deposition rates were to remain constant at the 2010 level in the future. However, it should be pointed out here that the deposition data used for comparison, which were determined on the basis of models throughout Germany [18], could be significantly higher locally, e.g., on areas with industrial and/or traffic concentrations or in the vicinity of a particularly high-emission plant.

3.5.1. Hg

Since no Germany-wide deposition calculations including dry deposition for Hg are available [17], the assessment of the total atmospheric load situation can only be made on the basis of the EMEP deposition mapping [10,11] (Table 6).

Table 6. Comparison of the Hg deposition (in g ha^{-1} a^{-1}) in the year 2013 [10,11] with the critical loads (in g ha^{-1} a^{-1}) of the receptor surfaces in Germany.

Deposition Hg 2013			Critical Loads		
EMEP [10,11]	EMEP [10,11]	CL(Hg)$_{eco}$	CL(Hg)$_{eco}$	CL(Hg)$_{drink}$	CL(Hg)$_{drink}$
5. Perc.	95th Perc.	Min.	5.–95. Perc. (Median)	Min.	5.–95. Perc. (Median)
0.00	0.87	0	0.2–0.6 (0.4)	0.26	0.6–5.7 (3.2)

Based on the evaluation of the EMEP total deposition values in a 50 km × 50 km grid [10,11], it can be assumed that a proportion of Germany's ecosystem receptor areas, i.e., those with high to medium sensitivity, are contaminated by Hg inputs above the critical load. Sensitive ecosystem types include in particular the unused deciduous forests of dry, nutrient-poor sites such as dry oak forests in the arid regions of Brandenburg and Saxony-Anhalt or beech forests along the coasts. The CL for drinking water protection in Germany are also clearly exceeded in the worst case. This means that in the worst case, Hg can accumulate in the soil or groundwater as soon as the critical limits in soil and soil water have been reached. However, the buffer capacity of the humus-rich forest soils in particular is very high for Hg, so that actual damage to ecosystem compartments, even if CL are exceeded, may only be expected after decades or centuries.

3.5.2. Cd

The comparison of the Germany-wide raster maps of the total deposition of Cd in 2010 determined by Schaap et al. [18] with the CL determined in this study is given in Table 7.

Table 7. Comparison of the Cd deposition (in g ha^{-1} a^{-1}) with the assessment values (in g ha^{-1} a^{-1}) and statistical evaluation of the area share of protected receptor areas in Germany, 2010, against limit values exceeded.

Deposition Cd 2010		Critical Loads					
Schaap et al. [18]	Schaap et al. [18]	CL(Cd)$_{eco}$	CL(Cd)$_{eco}$	CL(Cd)$_{drink}$	CL(Cd)$_{drink}$	CL(Cd)$_{food}$	CL(Cd)$_{food}$
5. Perc.	95. Perc.	Min.	5.–95. Perc. (Median)	Min.	5.–95. Perc. (Median)	Min.	5.–95. Perc. (Median)
0.21–0.45 (0.29)	2.33	1.53	4.1–42.4 (10.5)	0.65	2.5–18 (10.2)	2.31	3–9.3 (6)

The maximum atmospheric deposition in 2010 may have exceeded the CL for the protection of ecosystems, drinking water, and food in a few areas if maximum deposition rates hit areas with very low CL (<2.3 g ha^{-1} a^{-1}). This means that in the worst case an unacceptable accumulation of Cd in soil and drinking water could occur as long as the deposition is above the respective critical load and the critical limits have already been reached. However, when the inputs from atmospheric deposition and fertilisers are determined summarily, there is a risk that arable land will be exceeded if maximum total inputs are applied to high to average sensitive soils.

The comparison of the maximum inputs of Cd by fertilisation with CL(Cd)$_{food}$ shows that risks to human health from the consumption of wheat products of German origin cannot be excluded, since an intolerable increase in the Cd content in the soil can at the same time lead to an intolerable increase in the Cd content in the wheat grain [44].

3.5.3. Pb

The comparison of the Germany-wide raster maps of the total deposition of Pb in 2010 calculated by Schaap et al. [18] with the CL determined in this paper is shown in Table 8.

Table 8. Comparison of Pb deposition 2010 (in g ha^{-1} a^{-1}) with the Critical Loads (in g ha^{-1} a^{-1}) of receptor surfaces in Germany.

Deposition Pb 2010		Critical Loads			
Schaap et al. [18]	Schaap et al. [18]	CL(Pb)$_{eco}$	CL(Pb)$_{eco}$	CL(Pb)$_{drink}$	CL(Pb)$_{drink}$
5th Perc.	95th Perc	Min	5–95 Perc. (median)	Min	5–95 Perc. (median)
4.43–11 (6.71)	87.25	1.97	6–601 (21)	2.8	9–61 (35)

The maximum atmospheric deposition in 2010 has exceeded the CL for the protection of ecosystems and drinking water in a large part of the areas. This means that an unacceptable accumulation of Pb in soil and especially in drinking water occurs as soon as the critical limits in soil and soil water are reached. When the inputs from atmospheric deposition and fertilization are determined summarily, there is an excess risk for all arable land and grassland if maximum total inputs are applied to high to medium sensitive soils.

3.5.4. As

Table 9 compares the grid maps of the total deposition of As in 2010 calculated throughout Germany by Schaap et al. [18] with the CL determined in this study.

Table 9. Comparison of As deposition in 2010 (in g ha^{-1} a^{-1}) with the CL (in g ha^{-1} a^{-1}) of receptor surfaces in Germany.

Deposition As 2010		Critical Loads			
Schaap et al. [18]	Schaap et al. [18]	CL(As)$_{eco}$	CL(As)$_{eco}$	CL(As)$_{drink}$	CL(As)$_{drink}$
5. Perc.	95. Perc	Min.	5.–95. Perc. (Median)	Min.	5.–95. Perc. (Median)
0.28–0.6 (0.38)	1.03	115	181–711 (414)	2	6–56 (31)

Even in the worst case (maximum deposition meets minimum CL), the CL for the protection of ecosystems and drinking water are clearly undercut. Even if the fertiliser inputs in the maximum are added to the total atmospheric inputs in the maximum, the CL for ecosystem or drinking water protection are not exceeded. If, in the critical load calculation for ecosystem protection, the minimum threshold [46] used instead of the critical concentration in the leachate according to Doyle et al. [47] (cited in: [48]), a minimum critical load of 2.9 g ha^{-1} a^{-1} would be obtained. In the worst-case scenario, this minimum would also not be exceeded by the maximum deposition. The same applies to the alternatively calculated minimum critical load for drinking water protection.

3.5.5. Cu

The comparison of the grid maps of the total deposition of Cu in 2010 calculated by Schaap et al. [18] throughout Germany with the CL determined in this paper is shown in Table 10.

Table 10. Comparison of the 2010 Cu deposition (in g ha^{-1} a^{-1}) with the critical loads (in g ha^{-1} a^{-1}) of the receptor surfaces in Germany.

Deposition Cu 2010		Critical Loads			
Schaap et al. [18]	Schaap et al. [18]	CL(Cu)$_{eco}$	CL(Cu)$_{eco}$	CL(Cu)$_{drink}$	CL(Cu)$_{drink}$
5th Perc.	95th Perc	Min	5–95 Perc. (median)	Min	5–95 Perc. (median)
3.1–10.67 (5.89)	29.42	7	13–710 (74)	484	1070–11,268 (6172)

The maximum atmospheric deposition in 2010 [18] has exceeded the CL for ecosystem protection on part of the areas. This means that there will be an unacceptable accumulation of Cu in soil and/or groundwater as soon as the critical limits in soil are exceeded. When the inputs from atmospheric deposition and fertilisation are determined summarily, there is an exceedance risk for all arable land and grassland when maximum total inputs meet high to average sensitive soils.

3.5.6. Zn

The comparison of the grid maps of the total deposition of Zn in 2010 calculated by Schaap et al. [18] throughout Germany with the CL determined in this paper is shown in Table 11.

Table 11. Comparison of the Zn deposition 2010 (in g ha^{-1} a^{-1}) with the critical loads (in g ha^{-1} a^{-1}) of the receptor surfaces in Germany.

Deposition Zn 2010		Critical Loads			
Schaap et al. [18]	Schaap et al. [18]	CL(Zn)$_{eco}$	CL(Zn)$_{eco}$	CL(Zn)$_{drink}$	CL(Zn)$_{drink}$
5. Perc.	95. Perc.	Min.	5.–95. Perc. (Median)	Min.	5.–95. Perc. (Median)
11.89–33.38 (19.08)	76.63	81	189–1032 (565)	1234	2848–28,316 (15,628)

Even in the worst case (maximum deposition meets minimum CL), the CL for the protection of ecosystems and drinking water are barely undershot. The minimum inputs of Zn with fertilisers are already so high that there is a high risk of the CL for ecosystem protection being exceeded for all arable land and grassland. Since limit concentrations for Zn in drinking water are not specified, a critical load for drinking water protection cannot be determined.

3.5.7. Cr

The comparison of the grid maps of the total deposition of Cr in 2010 calculated by Schaap et al. [18] throughout Germany with the CL determined in this paper is shown in Table 12.

Table 12. Comparison of Cr deposition 2010 (in g ha^{-1} a^{-1}) and critical loads (in g ha^{-1} a^{-1}) of receptor areas in Germany.

Deposition 2010		Critical Loads			
Schaap et al. [18]	Schaap et al. [18]	CL(Cr)$_{eco}$	CL(Cr)$_{eco}$	CL(Cr)$_{drink}$	CL(Cr)$_{drink}$
5. Perc.	95. Perc.	Min.	5.–95. Perc. (Median)	Min.	5.–95. Perc. (Median)
0.84–2.08 (1.22)	3.97	78	115–448 (263)	12	28–282 (156)

The maximum atmospheric deposition in 2010 may not have exceeded the CL for the protection of ecosystems and drinking water in any area. This means that in the worst case there is no unacceptable accumulation of Cr in soil and drinking water. If, in the critical load calculation for ecosystem protection, instead of the critical concentration in the leachate according to Crommentuijn et al. [49] (cited in: [48]), the minimum threshold [46] was used as an alternative, a minimum critical load of 2.1 g ha^{-1} a^{-1} would be obtained. In the worst case, however, this minimum would be exceeded by the maximum deposition. The same applies to the alternatively calculated minimum critical load for drinking water protection. There is also an exceedance risk for a part of the arable land in Germany when the entries from atmospheric deposition and fertilisation are determined summarily, if maximum total entries hit soils with high to average sensitivity.

3.5.8. Ni

Table 13 compares the grid maps of the total deposition of Ni in 2010 calculated by [18] throughout Germany with the CL determined in this study.

Table 13. Comparison of Ni deposition in 2010 (in g ha^{-1} a^{-1}) with the critical loads (in g ha^{-1} a^{-1}) of receptor areas in Germany.

Deposition 2010		Critical Loads	
Schaap et al. [18]	Schaap et al. [18]	CL(Ni)$_{eco}$	CL(Ni)$_{eco}$
5. Perc.	95. Perc.	Min.	5.–95. Perc. (Median)
1.98–3.92 (2.69)	7.11	37	109–3338 (518)

The maximum atmospheric deposition in 2010 may not have exceeded the CL for ecosystem protection in any area. This means that in the worst case there is no unacceptable accumulation of Ni in the soil. Since limit concentrations for Ni in drinking water are not specified, a critical load for drinking water protection cannot be determined. There is no risk that the CL for the protection of forest, arable and grassland ecosystems will be exceeded when the inputs from atmospheric deposition and fertilisation are summarily determined, even if maximum total inputs would occur on high to average sensitive soils. If instead of the WHAM modelling results [50] for the critical concentration in leachate, the critical load calculation for ecosystem protection were to alternatively use the minimum threshold [46], a minimum critical load of 7.9 g ha^{-1} a^{-1} would be obtained. In the worst case, this minimum would also not be exceeded by the maximum deposition. The same applies to the alternatively calculated minimum critical load for drinking water protection.

3.5.9. Tl

CL for Tl for the protection of ecosystems cannot yet be determined, because there is no valid database for the derivation of impact-based ecosystem critical limits. A provisional rough estimate of the risk of Tl inputs into Germany's receptor ecosystems can be based on a calculated balance of inputs and outputs in the ranges typical for Germany (Table 14).

Table 14. Calculation of the acceptable Tl discharge in the typical German range (minimum and maximum).

Terms of the Balance Sheet	Acre	Grassland	Forest
Yield minimum (t dry mass ha^{-1} a^{-1})	2.199	0.1	0.65
Yield maximum (t dry mass ha^{-1} a^{-1})	14.088	6.5	7.4
Tl withdrawal by biomass harvest Minimum (g ha^{-1} a^{-1})	0.11	0.005	0.033
Tl removal by biomass harvest maximum (g ha^{-1} a^{-1})	0.704	0.325	0.37
Leakage water rate minimum (m^3 ha^{-1} a^{-1})	175	125	70
Seepage water rate maximum (m^3 ha^{-1} a^{-1})	949	678	380
Acceptable Tl- washing rate minimum (g ha^{-1} a^{-1})	0.035	0.025	0.014
Acceptable Tl- washout rate maximum (g ha^{-1} a-1)	0.19	0.136	0.076
Acceptable total Tl discharge minimum (g ha^{-1} a^{-1})	0.145	0.03	0.046
Acceptable total Tl discharge maximum (g ha^{-1} a^{-1})	0.894	0.46	0.446

The average Tl content in vegetable biomass on unpolluted soils is 0.05 g t^{-1} dry matter [51]. Thus, the range of Tl outputs with the biomass harvest is obtained by multiplying this typical concentration in plant stands of unpolluted soils by the minimum and maximum yields in Germany. In leachate, the minimum threshold of 0.0002 g m^{-3} [46] may be used as a critical limit. This threshold value is an ecotoxicologically determined threshold value for a Tl limit concentration.

Since no nationwide deposition surveys are available, no assessment of the pollution situation can be made. However, a measuring station in Dortmund, for example, recorded an annual average concentration in 2013, which converted into a deposition rate of 0.15 g ha^{-1} a^{-1} [52]. This value is within the range for the acceptable discharge rate.

Thus, a risk cannot be excluded that sensitive ecosystems could be overburdened in the long term.

3.5.10. V

CL for V for the protection of ecosystems cannot yet be determined because there is no valid database for the derivation of impact-based ecosystem-critical limits. A provisional rough estimate of the risk of V inputs into Germany's receptor ecosystems can be based on a calculated balance of inputs and outputs in the ranges typical for Germany (Table 15).

Table 15. Calculation of the acceptable V discharge in the typical German range (minimum and maximum).

Terms of the Balance Sheet	Acre	Grassland	Wood
Yield minimum (t dry mass ha^{-1} a^{-1})	2.199	0.1	0.65
Yield maximum (t dry mass ha^{-1} a^{-1})	14.088	6.5	7.4
V extraction by biomass harvest minimum (g ha^{-1} a^{-1})	1.54	0.07	0.455
V extraction by biomass harvest maximum (g ha^{-1} a^{-1})	9.86	4.55	5.18
Leakage water rate minimum (m^3 ha^{-1} a^{-1})	175	125	70
Seepage water rate maximum (m^3 ha^{-1} a^{-1})	949	678	380
Acceptable V-washing rate minimum (g ha^{-1} a^{-1})	0.7	0.5	0.28
Acceptable V-washing rate minimum (g ha^{-1} a^{-1})	3.8	2.71	1.52
Acceptable total V discharge minimum (g ha^{-1} a^{-1})	2.24	0.57	0.735
Acceptable total V discharge minimum (g ha^{-1} a^{-1})	13.66	7.62	6.697

Biomass harvesting is one of acceptable cutting methods. The average V content in vegetable biomass can be assumed to be 0.7 g t^{-1} dry matter [53]. The ranges of the V outputs with the biomass harvest then result from multiplying this typical concentration in plant stands of unpolluted soils with the minimum and maximum yields in Germany. Furthermore, leaching with the seepage can be taken into account as an acceptable discharge, whereby the critical concentration of the metal in the leachate can be assumed to be the negligibility threshold of 0.004 g m^{-3} [46]. This human-toxic threshold value is lower than the ecotoxicological threshold value for a V limit concentration.

The maximum deposition 2010 [18] for forest and grassland is slightly above the minimum acceptable deposition. In the worst case (maximum deposition meets areas with minimal cut-offs), a risk of impairment of ecosystem functions cannot be ruled out.

3.6. Comparison and Discussion of Assessment Values, Risk Assessment of Heavy Metal Inputs

When comparing the deposition calculated by Schaap et al. [18] as an area-wide dataset for Germany with the assessment values, the differences between the calculation results of the deposition (with EMEP model, LOTO-EUROS model, derived from mosses) and between assessment values are of great importance. The dataset for MH deposition calculated for Germany corresponds methodically roughly to the German dataset for the total deposition of N, which is used for the assessment of environmental impacts of projects and plants as the data basis for determining the background deposition. However, such an application is not recommended for the datasets of the total heavy metal deposition due to the high uncertainties [18]. However, it is in line with expectations that this background deposition will not exceed assessment values for the assessment of the environmental impacts of plants or projects.

3.6.1. Protection of Human Health

Comparing the assessment values based on human toxicological thresholds and relating to the total general exposure with the corresponding airborne input rates (for Hg:

EMEP deposition grid data for 2013; for all other metals deposition grid data for 2010 from Schaap et al. [18], then the following undercuttings and exceedances become obvious (Table 16). A comparison of the plant-related assessment values for deposition according to TA Luft [36] with the background loads does not provide any information on the currently existing risks and is therefore not made in the following section. Alternatively, it is indicated to what extent the background deposition already exhausts the values according to Table 6 or Table 8 of the TA Luft [36] or comparable assessment values.

Table 16. Assessment values for heavy metal (in g ha^{-1} a^{-1}) fluxes for the protection of human health and their exceedance/undercutting by atmospheric deposition (for Hg: EMEP in 2013 [10,11]; for all others German dataset in 2010 from Schaap et al. [18]).

Metal	TALuft Table 6	TALuft Table 8	39th BImSchV Coniferous/Deciduous Forest/ Arable Land [1]	EU-Position Paper Coniferous/Deciduous Forest/ Arable Land [1]	CL(M)$_{Food}$	CL(M)$_{Drink}$
	Emitter-Related			**General Load**		
Hg	4[+]	110[+]				0.3–13.8[−−]
Cd	7[+]	9[+]	7/4/2.5[+]	9–18[+]	1.9–19.2[−]	0.8–42.6[−]
Pb	365[+]	675[+]	716/420/250[+]			3–142[−−]
As	15[+]	4271[+]	6/4/2.2[+]	3–9/4–13/1.5–5[+]		2–138[+]
Ni	55[+]		28/17/10[+]	8–42/14–72/5–25[−]		
Cu						484–27,533[+]
Zn						1234–69,133[+]
Cr						12–688[+]

[1] Converted into input rates using the mean deposition velocities in the various vegetation complexes. [+] Critical values for human health are not exceeded. A risk can be excluded. [−] Critical values for human health are exceeded in worst cases. A risk cannot be excluded regionally. [−−] Critical values for human health are exceeded. There is a regional risk.

Although the immission limit values and target values of the 39th Federal Immission Control Ordinance [38] are in principle suitable as assessment values for endangering human health, the concentrations stated are not directly comparable with the deposition of the German dataset. If the concentrations are converted into input rates using the mean deposition velocities in the various vegetation complexes, taking into account the proportions of coarse and fine fractions in the dust, different permissible input rates are obtained for coniferous and deciduous forest and for arable land.

The EU Position Paper [54] specifies immission target values (concentrations in the particulate matter fraction PM10) for As, Cd, and Ni. For Cd, a deposition threshold derived from concentration values is also proposed. These are proposals derived from human toxicological data. These assessment values are therefore suitable for the risk assessment of total entries for human health. The assessment concentrations for As and Ni were converted into allowable input rates as indicated above. For the protection of drinking water, the CL (CL(M)$_{drink}$) for the atmospheric total heavy metal inputs for the German receptor surfaces on a scale of 1:1 million were determined as assessment values for the assessment of the CL(M)$_{drink}$, in which the limit concentrations from the German Drinking Water Ordinance [55] were included as critical threshold values. These are identical to the corresponding limit concentrations of the WHO guideline [56]. The CL for drinking water protection were determined taking into account the different leachate rates and vegetation types. In this respect, they show a higher degree of differentiation than the absolute assessment values of the 39th BImSchV [38] and the TA Luft [36]. Entries at the CL level lead to an equilibrium between total input and unpolluted output and thus guarantee the precautionary avoidance of an accumulation of HM in drinking water above the limit values. Thus, the following differentiated picture emerges with regard to the risk to human health from inputs of the individual HM under consideration.

Hg

The 39th BImSchV [38] and the EU Position Paper [54] do not contain assessment values for Hg. Based on the EMEP deposition grid map of Germany for the year 2013, the CL for drinking water protection CL(Hg) drink were exceeded in the German dataset 2016 [34] with a focus on southeast Brandenburg, northeast Saxony, and southwest Saxony-Anhalt. The maximum deposition exhausts the assessment value of TA Luft [36] (Table 6) to 22% and the assessment value of TA Luft [36] (Table 8) to 0.8%.

Cd

For the protection of plant food (wheat grain), a critical load for Cd inputs on wheat fields (CL(Cd)$_{food}$) was determined as the assessment value, in which the Cd limit concentration for wheat was included in accordance with the recommendation of the manual [8,57]. This is half of the limit value set in the EU regulation (EC No. 1881/2006). The CL(Cd)$_{food}$ was not exceeded in 2010 by the atmospheric Cd deposition in the receptor surfaces of the German CL dataset 2016 [34]. However, it should be noted that only a fraction of the heavy metal load on agricultural soils results from the atmosphere. In particular, the comparison of Cd inputs by fertilisation with CL(M)$_{food}$ shows the risk of harmful accumulation in wheat fields. Since the current content of Cd in wheat correlates with the content in soil [44], there is currently a risk potential for human health from the consumption of wheat products of German origin. The CL for drinking water protection in the receptor areas of the German dataset 2016 [34] are not exceeded by the atmospheric deposition in 2010 (Table 4). In the German dataset 2016 [34], it may be possible that areas where maximum deposition rates meet a very low critical load (<2.3 g ha^{-1} a^{-1}) (worst case) may not be mapped due to scale conditions. In these cases, the maximum atmospheric deposition in 2010 (Table 4) may have exceeded the CL for the protection of drinking water and food. This means that in the worst case, Cd may accumulate in drinking water and wheat products as long as the deposition is above the respective critical load and the critical limits are exceeded.

The CL(Cd)$_{drink}$ and CL(Cd)$_{food}$ are predominantly in the range of the range from the EU position paper [54], but also go deeper than this. The target value for the Cd entry from the EU position paper is far below the atmospheric deposition 2010 [18]. The maximum deposition exhausts the assessment value of TA Luft [36] (Table 6) at 27% and the assessment value of TA Luft [36] (Table 8) at 21%. If the assessment value for the concentration from the 39th BImSchV [38] were alternatively converted into a deposition, this would result in assessment values of 2.5–7 g ha^{-1} a^{-1}, which would not be exceeded by the deposition in the German dataset or in the worst case. However, this is only an auxiliary calculation for a rough comparison of the exceedance rates of CL(M)$_{drink}$ and CL(M)$_{food}$.

Pb

The CL for the protection of drinking water will be exceeded by atmospheric Pb deposition in 2010 [18] on 2.41% of the receptor areas in Germany, predominantly in the state of Brandenburg, in Leipzig and in the Ruhr area. This area proportion may be higher, since the German dataset 2016 [34] may not include areas where maximum deposition rates meet a very low critical load (worst case) (Table 8). The maximum deposition [18] exhausts 19% of the assessment value of the TA Luft [36] (Table 6) and 11% of the assessment value of the TA Luft [36] (Table 8). If the assessment value for the concentration from the 39th BImSchV [38] were alternatively converted into a deposition rate, this would result in assessment values of 250−716 g ha^{-1} a^{-1}, which are not exceeded by the deposition of the German dataset [18] and also in the worst case.

As

The CLs for drinking water protection are not exceeded in the receptor areas of the German dataset 2016 [34] by the atmospheric deposition in 2010 [18]. Even in the worst case (maximum deposition rates meet the lowest critical load), which does not occur in the German dataset 2016 [34] but could occur on a larger scale, exceeding the CL for drinking water protection is ruled out. Even if a critical load calculation for drinking water protection is carried out alternatively on the basis of the minimum threshold value [46], the resulting minimum critical load in the worst case would not be exceeded by the maximum deposition. The Minority threshold [46] corresponds to the base value.

The maximum deposition exhausts 6% of the assessment value of the TA Luft [36] (Table 6) and 0.02% of the assessment value of the TA Luft [36] (Table 8). If the assessment value for the concentration from the 39th BImSchV [38] were alternatively converted into a deposition, this would result in assessment values of 2.5–6 g ha^{-1} a^{-1}, which are not exceeded by the deposition of the German dataset and also in the worst case.

Cu

The CL for drinking water protection in the receptor areas of the German dataset 2016 [34] are not exceeded by the atmospheric deposition in 2010 [18]. Even in the worst case (maximum deposition rates meet the lowest critical load), which does not occur in the German dataset 2016 [34], but could occur on a larger scale, exceeding the CL for drinking water protection is ruled out.

Zn

The CL for drinking water protection in the receptor areas of the German dataset 2016 [34] are not exceeded by the atmospheric deposition in 2010 [18]. Even in the worst case (maximum deposition rates meet the lowest critical load), which does not occur in the German dataset 2016 [34] but could occur on a larger scale, exceeding the CL for drinking water protection is ruled out.

Cr

The CL for drinking water protection in the receptor areas of the German dataset 2016 [34] are not exceeded by the atmospheric deposition in 2010 [18]. Even in the worst case (maximum deposition rates meet the lowest critical load), which does not occur in the German dataset 2016 [34] but could occur on a larger scale, exceeding the CL for drinking water protection is ruled out. However, if a critical load calculation for drinking water protection was carried out alternatively on the basis of the minimum threshold value [46], the resulting minimum critical load would be exceeded by the maximum deposition [18] in the worst case.

Ni

The BTrinkwV [55] does not specify a limit concentration for Ni, therefore no critical load for drinking water protection is calculated in this study. If a critical load calculation for drinking water protection is carried out alternatively on the basis of the minimum threshold value [46], the resulting minimum critical load in the worst case would not be exceeded by the maximum deposition. The maximum deposition exhausts 12% of the TA Luft [36] (Table 6) assessment value.

If the assessment value for the concentration from the 39th BImSchV [38] were alternatively converted into a deposition, this would result in assessment values of 10–28 g ha^{-1} a^{-1}, which would not be exceeded by the deposition of the German dataset and also in the worst case. If the target values of the EU position paper [54] are converted into deposition rates (5–72 g ha^{-1} a^{-1}), in the worst case (maximum deposition meets minimum permissible input) there is an exceedance risk on arable land and grassland.

Tl

No limit concentration for Tl is specified in the BTrinkwV [55]. Therefore, no critical load for the drinking water protection $CL(Tl)_{drink}$ was calculated in this study. A human toxicological minimum threshold is also not available [46]. An exceedance of the immission values for heavy metal deposition according to TA Luft [36] (Tables 6 and 8) by the diffuse total pollution from the long-distance transport of the metal in the atmosphere cannot be determined nationwide. However, the assessment values are far above, for example, the annual average concentration measurement values 2013 of the LANUV NRW [52] in Dortmund converted into deposition rates.

V

The BTrinkwV [55] does not specify a limit concentration for V. Therefore, no critical load for the drinking water protection $CL(Ni)_{drink}$ is calculated in this study. A risk assessment using an input/output balance based on a human toxicological minimum threshold [46] shows that a health risk cannot be safely excluded by the inputs in 2010.

3.6.2. Protection of Terrestrial Ecosystems (in Particular Soils) from Harmful Changes

If one compares the assessment values with the corresponding aerial input rates (for Hg: EMEP deposition grid data for 2013 [10,11]; for all other metals deposition grid data for 2010 according to [18], the following under- and overruns result (Table 17). However, the assessment of risks from airborne inputs on the basis of the assessment values from the various statutory regulations and recommendations must be considered taking into account different levels of protection, protection objectives, and impact thresholds. The assessment values of the TA Luft [36], the 39th BImSchV [38], and the EU Position Paper [54] are based on human toxicological threshold values but are also intended for the protection of plants and the environment in general, whereby it is assumed that the ecosystem compartments are not more sensitive than humans. After conversion of the permissible concentrations from the 39th BImSchV [38] into permissible annual input rates, there are no exceedances in 2010 due to the atmospheric background deposition [18]. The assessment values from the EU position paper [54] for Cd will not be exceeded in 2010, as will the assessment values for As after conversion. The converted lowest assessment values for Ni are exceeded in the worst case on fields and grassland.

Table 17. Assessment values for heavy metal fluxes to protect ecosystems and their exceedance/undercutting by atmospheric deposition (for Hg: EMEP in 2013 [10,11]; for all others German dataset in 2010 [18]).

	Assessment Value (LUA Brandenburg 2008)	TALuft Table 6	TALuft Table 8	BBodSchV Permissible Additional Load	39th BImSchV Coniferous/Deciduous Forest/Arable Land [1]	EU-Position Paper Coniferous/Deciduous Forest/Arable Land [1])	CL(M)eco
	Emittent-Related				General Load		
	$(\text{g ha}^{-1}\ 100\ \text{a}^{-1})$				$(\text{g ha}^{-1}\ \text{a}^{-1})$		
Hg	2[+]	4[+]	110[+]	1.5[+]			0.1–1.1[--]
Cd	5[+]	7[+]	9[+]	6[+]	7/4/2.5[+]	9–18[+]	1.5–127.6[-]
Pb	768[+]	365[+]	675[+]	400[+]	716/420/250[+]		2–2603[--]
As		15[+]	4271[+]		6/4/2.2[+]	3–9/4–13/1.5–5[+]	115–1669[+]
Ni	154[+]	55[+]		100[+]	28/17/10[+]	8–42/14–72/5–25[-]	37–11,232[+]
Cu				360[+]			7–3384[--]
Zn				1200[+]			81–2457[+]
Cr				300[+]			78–1049[+]

[1] Converted into input rates using the mean deposition velocities in the various vegetation complexes. [+] Critical values for ecosystems are not exceeded. A risk can be excluded. [-] Critical values for ecosystems are exceeded in worst cases. A risk cannot be excluded regionally. [--] Critical values for ecosystems are exceeded. There is a regional risk.

At the same time, a comparison of the plant-related assessment values for metal inputs (assessment value according to LUA Brandenburg [58], TA Luft [36], permissible additional load according to BBodSchV [37]) with the background loads does not provide any information on the currently existing spatially widespread risks for ecosystems. The comparison can only serve to roughly estimate to what extent the permissible total load has already been exhausted by the background load. The uncertainties of the deposition mapping and its small scale do not permit concrete spatial statements for individual sites. It was therefore to be expected that the atmospheric entries for 2010 would not exceed the plant-related assessment values.

In the following, only the risks of terrestrial ecosystems, including soil, are discussed on the basis of assessment values based on ecotoxicological thresholds, i.e., from the BBodSchV [37], the Brandenburg enforcement aid for the FFH-habitats impact assessment [58] and the CL for ecosystem protection identified in Schröder et al. [34]. As the pathways of action of HM to all compartments of a terrestrial ecosystem usually run across the soil, the assessment values for the protection of soil functions can at the same time be regarded as relevant assessment values for the protection of ecosystems.

The CL for ecosystem protection [34] for the heavy metal inputs of Hg, Cd, Pb, As, Cu, Zn, Cr, and Ni ($CL(M)_{eco}$) are differentiated by ecosystem type. The input values are determined soil- and vegetation-specifically. The critical threshold values (critical concentration in the soil for the maintenance of microbial functions, for the protection of plants, invertebrates, and soil microorganisms) are comparable with the criteria for determining the precautionary values in the Federal Soil Protection Ordinance [37]. However, the precautionary values of the BBodSchV [37] are stated as concentrations and are therefore not directly comparable with the flow rates of the CL.

According to current knowledge, compliance with $CL(M)_{eco}$ with the inclusion of all input paths permanently (for all time) excludes the possibility of risks arising, provided that the critical limits (critical concentrations in the indicators considered) have not yet been exceeded. If they have already been exceeded, compliance with the critical load leads to a long-term, gradual reduction up to the critical limits.

The assessment values of the ´Brandenburgische FFH-Vollzugshilfe´ [58] are given as maximum permissible concentrations in the soil. In addition, plant-related irrelevance thresholds can be calculated. A comparison of the permissible enrichment rate in 100 years with background deposition therefore also does not lead to a real risk assessment. In the following, therefore, only the extent to which the total permissible positions calculated in this way are already exhausted by the entries at background level is given.

In the following, an assessment of the risk of harmful changes in terrestrial ecosystems based on the current 2010 and 2013 HM deposition under consideration is nevertheless to be carried out, taking into account the limited comparability of the assessment values.

Hg

On the basis of the EMEP deposition grid map of Germany for 2013 [10], the CLs for ecosystem protection in the German dataset 2016 [34] are exceeded, with a focus on North Rhine-Westphalia, southeast Brandenburg, north-east Saxony, and southwest Saxony-Anhalt.

In addition, further diffuse Hg entries may have to be taken into account, which aggravates the situation. However, due to the high buffering capacity of the soils for Hg, it cannot be concluded from a $CL(Hg)_{eco}$ exceedance that there is an immediate risk.

The permissible annual additional load of Hg according to BBodSchV [37] in the amount of 1.5 g ha^{-1} a^{-1}, if the precautionary value of the Hg concentration in the soil according to BBodSchV [37] has already been reached or exceeded, is far above the maximum of the EMEP deposition in 2013 [10,11]. Fifty-eight percent of the maximum atmospheric background deposition uses this value, but the permissible additional load also applies to all other input paths in total. The irrelevant plant-related additional load extrapolated from the irrelevance threshold according to LUA Bbg [58] using the example of a cambisol to 100 years is already 44% exhausted by the maximum atmospheric background deposition in 2013 [18].

Cd

The CL for ecosystem protection in the receptor areas of the German dataset 2016 [34] are not exceeded by the atmospheric deposition 2010 [18]. In the German dataset 2016 [34], it may be possible that areas where maximum deposition rates meet a very low critical load ($<2.3\,\mathrm{g\,ha^{-1}\,a^{-1}}$) (worst case) may not be depicted due to scale. In these cases, the maximum atmospheric deposition in 2010 may have exceeded the CL for ecosystem protection. The permissible annual Cd input rate according to BBodSchV [37] of $6\,\mathrm{g\,ha^{-1}\,a^{-1}}$), if the precautionary value (Cd concentration in soil) according to BBodSchV [37] has already been reached or exceeded, is far above the maximum of the EMEP deposition in 2013 [10,11]. Thirty-nine percent of the maximum atmospheric background deposition uses this value, but the permissible additional load applies to all input paths.

The irrelevant plant-related additional load extrapolated from the irrelevance threshold according to LUA Bbg [58] using the example of a cambisol to 100 years is already 47% exhausted by the maximum atmospheric background deposition in 2010 [18].

Pb

The CL for Pb inputs for the protection of ecosystems are exceeded by the atmospheric deposition in 2010 [18] on 14.11% of the receptor areas in Germany, predominantly in the Leipzig and Thuringian bight, in the Harz foreland and in the Ruhr area. This area share may be higher, since the German dataset 2016 [34] may not include areas where maximum deposition rates meet a very low critical load (worst case). The permissible annual Pb input rate under the BBodSchV [37] of $400\,\mathrm{g\,ha^{-1}\,a^{-1}}$, if the precautionary value has already been reached or exceeded, is well above the maximum of the 2010 deposition [18]. The maximum atmospheric background deposition exhausts 22% of this value, but the permissible additional load applies to all entry paths. The irrelevant plant-related additional load extrapolated from the irrelevance threshold according to LUA Bbg [58] using the example of a cambisol to 100 years is used up by 11% by the maximum atmospheric background deposition in 2010 [18].

As

The CL for ecosystem protection in the receptor areas of the German dataset 2016 [34] are not exceeded by the atmospheric deposition 2010 [18]. Even in the worst case (maximum deposition rates meet the lowest critical load), which does not occur in the German dataset 2016 [34] but could occur on a larger scale, exceeding the CL for ecosystem protection is ruled out. Even if a critical load calculation for ecosystem protection was carried out alternatively on the basis of the minimum threshold value [46], the resulting minimum critical load in the worst case would not be exceeded by the maximum deposition [18]. A permissible additional annual input rate of As according to BBodSchV [37] is not specified.

Cu

In 2010 [18], the Cu deposits exceeded the $\mathrm{CL(Cu)_{eco}}$ on 1.16% of the receptor areas in Germany, predominantly in the Berlin environs and the Ruhr area. This area proportion may be higher, since the German CL dataset 2016 [34] may not reflect areas where maximum deposition rates meet a very low critical load (worst case). The permissible annual input rate of Cu according to BBodSchV [37] of $360\,\mathrm{g\,ha^{-1}\,a^{-1}}$, if the permissible Cu concentration in the soil according to BBodSchV [37] has already been reached or exceeded, is far above the maximum of the deposition 2010 [18]. The maximum background deposition exhausts this value to 8%, but the permissible additional load applies to all input paths.

Zn

The CL for ecosystem protection in the receptor areas of the German data set 2016 [34] are not exceeded by the atmospheric deposition 2010 [18]. Even in the worst case (maximum deposition rates meet the lowest critical load), which does not occur in the German data set 2016 [34] but could occur on a larger scale, exceeding the CL for ecosystem protection is ruled out. The permissible annual input rate of Zn according to BBodSchV [37] of 1200 g ha^{-1} a^{-1}, if the permissible Zn concentration in the soil according to BBodSchV [37] has already been reached or exceeded, is far above the maximum of the deposition in 2010 [18]. The maximum background deposition exhausts this value by 6%, but the permissible additional load applies to all input paths.

Cr

The CL for ecosystem protection in the receptor areas of the German dataset 2016 [34] are not exceeded by the atmospheric deposition 2010 [18]. Even in the worst case (maximum deposition rates meet the lowest critical load), which does not occur in the German CL dataset 2016 [34] but could occur on a larger scale, exceeding the CL for ecosystem protection is ruled out. However, if a critical load calculation for ecosystem protection was carried out alternatively on the basis of the minimum threshold value [46], the resulting minimum critical load would be exceeded by the maximum deposition in the worst case. The permissible annual input rate of Cr according to BBodSchV [37] of 300 g ha^{-1} a^{-1}, if the permissible Cr concentration in the soil according to BBodSchV [37] has already been reached or exceeded, is far above the maximum of the deposition in 2010 [18]. The maximum background deposition exhausts this value by 1.3%, but the permissible additional load applies to all input paths.

Ni

The CL for ecosystem protection in the receptor areas of the German CL dataset 2016 [34] are not exceeded by the atmospheric deposition 2010 [18]. Even in the worst case (maximum deposition rates meet the lowest critical load), which does not occur in the German CL dataset 2016 [34], but could occur on a larger scale, exceeding the CL for ecosystem protection is ruled out. Even if a critical load calculation for ecosystem protection was carried out alternatively on the basis of the minimum threshold value [46], the resulting minimum critical load would not be exceeded by the maximum deposition [18] in the worst case. The permissible annual Ni input rate according to BBodSchV [37] of 100 g ha^{-1} a^{-1}, if the permissible Ni concentration in the soil according to BBodSchV [37] has already been reached or exceeded, is far above the maximum of the deposition in 2010 [18]. The maximum background deposition exhausts this value by 7%, but the permissible additional load applies to all input paths.

The maximum atmospheric background deposition in 2010 [18] exploits 4.6% of the irrelevance threshold according to LUA Bbg [58] extrapolated from the example of a cambisol to 100 years of irrelevant plant-related additional pollution.

Tl

Since no nationwide deposition surveys are available, no assessment of the pollution situation can be made. However, a measuring station in Dortmund, for example, recorded an annual average concentration in 2013, which converted to a deposition rate of 0.15 g ha^{-1} a^{-1} [52]. A risk assessment by means of an input/output balance based on an ecotoxicological minimum threshold [46] shows that a risk cannot be safely excluded, at least in the Ruhr region. Tl is a highly toxic element for living organisms, comparable to the effect of Hg [59]. Tl is hardly transported in the soil and is thus strongly enriched in the rooted topsoil during prolonged input [59,60] states that 80% of anthropogenic Tl is stored in humus-rich topsoil. This results in an obviously neglected need for research. The impact-based derivation of assessment values and at the same time the inventory of the current deposition in Germany are urgently required.

V

A risk assessment by means of an input/output balance based on a human toxicological minimum threshold [46], which is lower than the ecotoxicological threshold determined, results in a health risk from the inputs in 2010 in the worst case. For example, the maximum deposition in 2010 [18] (Table 4) for forest and grassland is slightly above the minimum acceptable deposition. In the worst case (maximum deposition meets areas with minimal outcropping), a long-term risk of impairment of ecosystem functions cannot be ruled out. Assessment values for V are not given in the BbodSchV [37].

4. Future Research

The German Moss Monitoring 2020 does not continue the monitoring network 2015 and the range of measured elements except for POPs. This is a cause for concern, because in that way it is not possible to detect a trend reversal like for Cr, Sb, Zn, and standstill of concentrations of Al, As, Cd, Cu, Hg, Ni, V between 2000 and 2005. The Moss Survey 2025 should therefore again focus on the internationally agreed measurement spectrum (Al, As, Cd, Cr, Cu, Fe, Hg, Ni, Pb, Sb, V, Zn; N; POP). From a political and ecological point of view, not only increases in concentrations are worth reporting, but also standstills and (further) declines.

Author Contributions: Conceptualization, S.N., A.S., and W.S.; methodology, A.S.; data curation, S.N. and A.S; writing—original draft preparation, W.S.; writing—review and editing, W.S.; supervision, W.S.; project administration, W.S.; funding acquisition, W.S. All authors have read and agreed to the published version of the manuscript.

Funding: This research was funded by the Federal Environment Agency, Germany.

Institutional Review Board Statement: Not applicable.

Informed Consent Statement: Not applicable.

Data Availability Statement: The provision of data is in preparation.

Acknowledgments: We would like to thank Gudrun Schütze (Federal Environment Agency) for her constructive professional support of the project.

Conflicts of Interest: The authors declare no conflict of interest.

References

1. Rutter, A.J. The hydrological cycle in vegetation. In *Vegetation and the Atmosphere*; Monteith, J.L., Ed.; Academic Press: London, UK, 1975; Volume 1, pp. 111–154.
2. Dollard, G.J.; Unsworth, M.H.; Harvey, M. Pollutant transfer in upland regions by occult precipitation. *Nature* **1983**, *302*, 241–243. [CrossRef]
3. Unsworth, M.; Wilshaw, J. Wet, occult and dry deposition of pollutants on forests. *Agric. For. Meteorol.* **1989**, *47*, 221–238. [CrossRef]
4. Bak, J.; Doytchinov, S.; Fisher, R.; Forsius, M.; Grennfelt, P.; Harmens, H.; Hettelingh, J.-P.; Holmberg, M.; Jenkins, A.; Krzyzanowski, M. *Impacts of Air Pollution on Ecosystems, Human Health and Materials under Different Gothenburg Protocol Scenarios*; Report by the Working Group on Effects; LRTAP: Geneva, Switzerland, 2012; pp. 1–45. Available online: https://unece.org/fileadmin/DAM/env/documents/2012/air/WGE_31th/N_2_WGE-impact-report-2012-sept.pdf. (accessed on 22 January 2021).
5. De Marco, A.; Proietti, C.; Anav, A.; Ciancarella, L.; D'Elia, I.; Fares, S.; Fornasier, M.F.; Fusaro, L.; Gualtieri, M.; Manes, F.; et al. Impacts of air pollution on human and ecosystem health, and implications for the National Emission Ceilings Directive: Insights from Italy. *Environ. Int.* **2019**, *125*, 320–333. [CrossRef] [PubMed]
6. Wright, L.P.; Zhang, L.; Cheng, M.-D.; Aherne, J.; Wentworth, G.R. Impacts and effects indicators of atmospheric deposition of major pollutants to various ecosystems—A Review. *Aerosol Air Qual. Res.* **2018**, *18*, 1953–1992. [CrossRef]
7. CLRTAP 2014. Guidance on Mapping Concentrations Levels and Deposition Levels, Manual on Methodologies and Criteria for Modelling and Mapping Critical Loads and Levels and Air Pollution Effects, Risks and Trends. UNECE Convention on Long-Range Transboundary Air Pollution. Available online: https://www.umweltbundesamt.de/en/manual-for-modelling-mapping-critical-loads-levels (accessed on 20 March 2015).

8. CLRTAP 2017. Revision of Guidance on Mapping Concentrations Levels and Deposition Levels, Manual on Methodologies and Criteria for Modelling and Mapping Critical Loads and Levels and Air Pollution Effects, Risks and Trends. UNECE Convention on Long-Range Transboundary Air Pollution. Available online: https://www.umweltbundesamt.de/en/manual-for-modelling-mapping-critical-loads-levels (accessed on 28 March 2017).

9. Colette, A.; Aas, W.; Banin, L.; Braban, C.F.; Ferm, M.; González Ortiz, A.; Ilyin, I.; Mar, K.; Pandolfi, M.; Putaud, J.-P. *EMEP Cooperative Programme for Monitoring and Evaluation of the Long-range Transmission of Air Pollutants in Europe. Air Pollution Trends in the EMEP Region between 1990 and 2012: Joint Report of EMEP Task Force on Measurements and Modelling (TFMM), Chemical Co-ordinating Centre (CCC), Meteorological Synthesizing Centre-East (MSC-e), Meteorological Synthesizing Centre-West (MSC-w)*; EMEP/CCC-report 1/2016; Norwegian Institute for Air Research: Kjeller, Norway, 2016; pp. 1–105.

10. EMEP 2015. EMEP Status Reports: Heavy Metals: Analysis of Long-Term Trends, Country-Specific Research and Progress in Mercury Regional and Global Modelling (No. 2/2015). Available online: http://www.msceast.org/index.php/reports (accessed on 20 July 2015).

11. Country Specific Information, Complementary to the EMEP Status Reports. Available online: http://www.msceast.org/index.php/emep-countries (accessed on 20 July 2015).

12. Engardt, M.; Simpson, D.; Schwikowski, M.; Granat, L. Deposition of sulphur and nitrogen in Europe 1900–2050. Model calculations and comparison to historical observations. *Tellus B Chem. Phys. Meteorol.* **2017**, *69*, 1328945. [CrossRef]

13. Ilyin, I.; Gusev, A.; Rozovskaya, O.; Strijkina, I. *Transboundary Pollution of Germany by Heavy Metals and Persistent Organic Pollutants in 2010*; EMEP/MSC-E J Technical Report 5/2012; EMEP/MSC-E: Moscow, Russia, 2012.

14. Ilyin, I.; Rozovskaya, O.; Travnikov, O.; Varygina, M.; Aas, W.; Uggerud, H.T. Heavy Metals: Transboundary Pollution of the Environment. J. EMEP Status Report 2/2011. Available online: http://www.msceast.org (accessed on 28 March 2017).

15. Ilyin, I.; Rozovskaya, O.; Travnikov, O.; Varygina, M.; Aas, W.; Uggerud, H.T. Heavy Metals: Transboundary Pollution of the Environment. J. EMEP Status Report 2/2013. Available online: http://www.msceast.org (accessed on 28 March 2017).

16. Schaap, M.; Roemer, M.; Sauter, F.; Boersen, G.; Timmermans, R.; Builjes, P.J.H.; Vermeulen, A.T. *LOTOS-Euros: Documentation*; TNO Report BandO-a R 2005/297; TNO Built Environment and Geosciences: Apeldoorn, The Netherlands, 2005.

17. Schaap, M.; Timmermans, R.M.; Roemer, M.; Boersen, G.; Builtjes, P.J.; Sauter, F.J.; Velders, G.J.; Beck, J.P. The LOTOS EUROS model: Description, validation and latest developments. *Int. J. Environ. Pollut.* **2008**, *32*, 270. [CrossRef]

18. Schaap, M.; Hendriks, C.; Jonkers, S.; Builtjes, P. Assessment of the atmospheric heavy metal deposition to terrestrial ecosystems in Germany. In *Auswirkungen der Schwer-Metall—Emissionen auf Luftqualität und Ökosysteme in Deutschland—Quellen, Transport, Eintrag, Gefährdungspotenzial [Impacts of Heavy Metal Emission on Air Quality and Ecosystems across Germany—Sources, Transport, Deposition and Potential Hazards]*; Final Report, Project No. (FKZ) 3713-63-253, Report No. (UBA-FB) 002635/E, UBA-TEXTE 106/2018; Schröder, W., Nickel, S., Schlutow, A., Nagel, H.-D., Scheuschner, T., Eds.; The German Environment Agency: Dessau, Germany, 2018; pp. 1–81.

19. Torseth, K.; Aas, W.; Breivik, K.; Fjæraa, A.M.; Fiebig, M.; Hjellbrekke, A.; Myhre, C.L.; Solberg, S.; Yttri, K.E. Introduction to the European Monitoring and Evaluation Programme (EMEP) and observed atmospheric composition change during 1972–2009. *Atmos. Chem. Phys. Discuss.* **2012**, *12*, 5447–5481. [CrossRef]

20. Schröder, W.; Holy, M.; Pesch, R.; Zechmeister, H.; Harmens, H.; Ilyin, I. Mapping atmospheric depositions of cadmium and lead in Germany based on EMEP deposition data and the European Moss Survey 2005. *Environ. Sci. Eur.* **2011**, *23*, 19. [CrossRef]

21. Schröder, W.; Pesch, R.; Harmens, H.; Fagerli, H.; Ilyin, I. Does spatial auto-correlation call for a revision of latest heavy metal and nitrogen deposition maps? *Environ. Sci. Eur.* **2012**, *24*, 20. [CrossRef]

22. Onianwa, P.C. Monitoring atmospheric metal pollution: A review of the use of mosses as indicators. *Environ. Monit. Assess.* **2001**, *71*, 13–50. [CrossRef] [PubMed]

23. Schröder, W.; Nickel, S. Mapping percentile statistics of element concentrations in moss specimens collected from 1990 to 2015 in forests throughout Germany. *Atmos. Environ.* **2018**, *190*, 161–168. [CrossRef]

24. Schröder, W.; Nickel, S.; Völksen, B.; Dreyer, A.; Wosniok, W. *Nutzung von Bioindikationsmethoden zur Bestimmung und Regional-isierung von Schadstoffeinträgen für eine Abschätzung des atmosphärischen Beitrags zu aktuellen Belastungen von Ökosystemen [Use of Bioindication Methods for the Determination and Regionalisation of Pollutant Inputs to Estimate the Atmospheric Contribution to Current Ecosystem Pressures]*; UBA-Texte 91/2019; Umwelt Bundesamt: Dessau, Germany, 2019; Volume 1–2, pp. 1–189, 1–296.

25. Berg, T.; Hjellbrekke, A.; Rühling, Å.; Steinnes, E.; Kubin, E.; Larsen, M.M.; Piispanen, J. *Absolute Deposition Maps of Heavy Metals for the Nordic Countries Based on the Moss Survey*; TemaNord 505; Nordic Council of Ministers: Copenhagen, Denmark, 2010; p. 35.

26. Berg, T.; Steinnes, E. Use of mosses (*Hylocomium splendens* and *Pleurozium schreberi*) as biomonitors of heavy metal deposition: From relative to absolute values. *Environ. Pollut.* **1998**, *98*, 61–71. [CrossRef]

27. Harmens, H.; Mills, G.; Hayes, F.; Norris, D.; Sharps, K. Twenty eight years of ICP Vegetation. An overview of its activities. *Ann. Bot.* **2015**, *5*, 31–43.

28. Schröder, W.; Hornsmann, I.; Pesch, R.; Schmidt, G.; Markert, B.; Fränzle, S.; Wünschmann, S.; Heidenreich, H. Nitrogen and metals in two regions in Central Europe: Significant differences in accumulation in mosses due to land use? *Environ. Monit. Assess.* **2007**, *133*, 495–505.

29. Schröder, W.; Holy, M.; Pesch, R.; Harmens, H.; Ilyin, I.; Steinnes, E.; Alber, R.; Aleksiayenak, Y.; Blum, O.; CosKun, M.; et al. Are cadmium, lead and mercury concentrations in mosses across Europe primarily determined by atmospheric deposition of these metals? *J. Soils Sediments* **2010**, *10*, 1572–1584. [CrossRef]

30. Herpin, U.; Lieth, H.; Markert, B. *Monitoring der Schwermetallbelastung in der Bundesrepublik Deutschland mit Hilfe von Moosanalysen [Monitoring of Heavy Metal Pollution in the Federal Republic of Germany Using Moss Analyses]*; UBA-Texte 31/95: Berlin, Germany, 1995; p. 161.

31. Siewers, U.; Herpin, U. *Schwermetalleinträge in Deutschland—Moosmonitoring 1995/96 [Heavy Metal Deposition in Germany—Moss monitoring 1995/96]*; Geologisches Jahrbuch Sonderhefte Sonderheft Heft 2; Schweizerbart: Stuttgard, Germany, 1998; p. 199.

32. Siewers, U.; Herpin, U.; Strassburg, S. *Schwermetalleinträge in Deutschland (Part II). Moosmonitoring 1995/96 [Heavy Metal Deposition in Germany (Part II). Moss Monitoring 1995/96]*; Geologisches Jahrbuch Sonderhefte Sonderheft Heft 3; Schweizerbart: Stuttgard, Germany, 2002; p. 103.

33. Pesch, R.; Schröder, W. Statistical and geoinformatical instruments for the optimisation of the German moss-monitoring network. In *Managing Environmental Knowledge, Proceedings of the 20th International Conference on Informatics for Environmental Protection, Graz, Austria, 6–8 September 2006*; Tochtermann, K., Scharl, A., Eds.; Shaker Verlag: Aachen, Germany, 2006; pp. 191–198.

34. Schröder, W.; Nickel, S.; Schlutow, A.; Nagel, H.-D.; Scheuschner, T. *Auswirkungen der Schwermetall—Emissionen auf Luftqualität und Ökosysteme in Deutschland—Quellen, Transport, Eintrag, Gefährdungspotenzial. Teil 2: Integrative Datenanalyse, Erheblichkeitsbeurteilung und Untersuchung der Gegenwärtigen Regelungen und Zielsetzungen in der Luftreinhaltung und Vergleich mit Ausgewählten Anforderungen, die sich in Bezug auf den Atmosphärischen Schadstoffeintrag aus den Verschiedenen Rechtsbereichen ergeben [Effects of Heavy Metal Emissions on Air Quality and Ecosystems in Germany—Sources, Transport, Input, Hazard Potential. Part 2: Integrative Data Analysis, Assessment of Relevance and Investigation of Current Regulations and Objectives in Air Pollution Control and Comparison with Selected Requirements Arising from the Various Areas of Legislation with Regard to Atmospheric Pollutant Input]*; UBA-TEXTE 107/2018: Dessau, Germany, 2018; pp. 1–257.

35. Nickel, S.; Schröder, W. Reorganisation of a long-term monitoring network using moss as biomonitor for atmospheric deposition in Germany. *Ecol. Indic.* **2017**, *76*, 194–206. [CrossRef]

36. Technische Anleitung zur Reinhaltung der Luft—TA Luft (Bundesministerium für Umwelt, Naturschutz und Reaktorsicherheit). Erste Allgemeine Verwaltungsvorschrift zum Bundes–Immissionsschutzgesetz. Available online: https://www.bmu.de/fileadmin/Daten_BMU/Download_PDF/Luft/taluft.pdf (accessed on 30 January 2021).

37. Bundes-Bodenschutz- und Altlastenverordnung (BBodSchV). (GBBl. I S. 1554 vom 12 Juli 1999, zuletzt geändert durch Art. 102 der Verordnung vom 31 August 2015 (GBl. I S. 1474) [Federal Soil Protection and Contaminated Sites Ordinance (BBodSchV). Available online: https://www.gesetze-im-internet.de/bbodschv/BBodSchV.pdf (accessed on 30 January 2021).

38. Neununddreißigste Verordnung zur Durchführung des Bundes-Immissionsschutzgesetzes Verordnung über Luftqualitätsstandards und Emissionshöchstmengen (39th BImSchV). [Thirty-Ninth Ordinance for the Implementation of the Federal Immission Control Act Ordinance on Air Quality Standards and Emission Ceilings]. Available online: https://www.gesetze-im-internet.de/bimschv_39/BJNR106510010.html (accessed on 30 January 2021).

39. *Directive 2004/107/EC of the European Parliament and of the Council of 15 December 2004 Relating to Arsenic, Cadmium, Mercury, Nickel and Polycyclic Aromatic Hydrocarbons in Ambient Air*; OJ L 23; European Parliament: Brussels, Belgium, 2005; pp. 3–16.

40. *Directive 2008/50/EC of the European Parliament and of the Council of 21 May 2008 on Ambient Air Quality and Cleaner air for Europe*; OJ L 152; European Parliament: Brussels, Belgium, 2008; pp. 1–44.

41. Nagel, H.-D.; Becker, R.; Kraft, P.; Schlutow, A.; Schütze, G.; Weigelt-Kirchner, R. *NFC Deutschland, Critical Loads, Biodiversität, Dynamische Modellierung [NFC Germany, Critical Loads, Biodiversity, Dynamic Modelling]*; UBA-TEXTE 39/2008: Dessau, Germany, 2008; pp. 1–291.

42. EP CLHM 2005. Expert Panel on Critical Loads of Heavy Metals (EP CLHM). How to Calculate Time Scales of Accumulation of Metals in Soils and Surface Waters? Some Methodological Principles. In *NFC Deutschland, Critical Loads, Biodiversität, Dynamische Modellierung [NFC Germany, Critical Loads, Biodiversity, Dynamic Modelling]*; Nagel, H.-D., Becker, R., Kraft, P., Schlutow, A., Schütze, G., Weigelt-Kirchner, R., Eds.; UBA-TEXTE 39/2008: Dessau, Germany, 2008; pp. 1–291.

43. NaSE. *Nationale Trendtabellen für die Deutsche Berichterstattung Atmosphärischer Emissionen (Schwermetalle) 1990–2015 [National Trend Tables for German Reporting of Atmospheric Emissions (Heavy Metals) 1990–2015]*; Stand Januar 2017; Umweltbundesamt: Dessau, Germany, 2017; Available online: https://www.umweltbundesamt.de/dokument/nationale-trendtabellen-fuer-die-deutsche-2 (accessed on 6 December 2017).

44. Knappe, F.; Möhler, S.; Ostermayer, A.; Lazar, S.; Kaufmann, C. Vergleichende Auswertung von Stoffeinträgen in Böden Über Verschiedene Eintragspfade [Comparative Evaluation of Substance Inputs into Soils via Different Input Paths]. Available online: https://www.umweltbundesamt.de/publikationen/vergleichende-auswertung-von-stoffeintraegen-in (accessed on 30 January 2021).

45. CLC 2006. Karte der Bodenbedeckung [Land Cover Map]. CORINE Land Cover 2006. Available online: https://www.umweltbundesamt.de/themen/boden-landwirtschaft/flaechensparen-boeden-landschaften-erhalten/corine-land-cover-clc (accessed on 28 March 2017).

46. Zeddel, A.; Quadflieg, A.; Utermann, J. Grundsätze für die Anwendung der Geringfügigkeitsschwellen an der Schnittstelle Wasserrecht—Abfallrecht—Bodenschutzrecht [Principles for the Application of Minimum Thresholds at the Interface of Water Law—Waste Law—Soil Protection Law]. Recht/Strategie/Wirtschaft 2016. Available online: https://www.vivis.de/wp-content/uploads/MNA3/2016_MNA_051-64_Zeddel.pdf (accessed on 28 March 2017).

47. Doyle, P.J.; Gutzman, D.W.; Sheppard, M.I.; Sheppard, S.C.; Bird, G.A.; Hrebenyk, D. An ecolocical risk assessment of air emissions of trace metals from copper and zinc production facilities. *J. Hum. Ecol. Risk Assess.* **2003**, *9*, 607–636. [CrossRef]

48. Reinds, G.J.; Groenenberg, J.E.; De Vries, W.D. *Critical loads of copper, nickel, zinc, arsenic, chromium and selenium for terrestrial ecosystems at a European scale. A preliminary assessment*; Alterra-Rapport; Alterra: Wageningen, The Netherlands, 2005; pp. 1–45.
49. Crommentuijn, T.; Polder, M.D.; van de Plassche, E.J. *Maximum Permissible Concentrations and Neglectible Concentrations for Metals, Taking Background Concentrations into Account*; Report Nr. 601501 001; National Institut of Public Health and Environment: Bilthoven, The Netherlands, 1997.
50. Tipping, E. WHAMC—A chemical equilibrium model and computer code for waters, sediments, and soils incorporating a discrete site/electrostatic model of ion-binding by humic substances. *J. Elsevier.* **1994**, *20*, 973–1023. [CrossRef]
51. Markert, B. Multi-element analysis in plant material-analytical tools and biological questions. In *Biochemistry of Trace Elements*; Adriano, D.C., Ed.; Lewis Publishers: Boca Raton, FL, USA, 1992; pp. 401–428.
52. LANUV NRW (Landesamt für Natur, Umwelt und Verbraucherschutz NRW). Jahreskenngrößen der LUQS-Messstetionen des LANUV NRW für die Jahre 2003–2013 [Annual characteristics of the LUQS measuring stations of LANUV NRW for the years 2003–2013]. 2015. Available online: https://www.lanuv.nrw.de/umwelt/luft/immissionen/messorte-und-werte (accessed on 28 March 2017).
53. Schroeder, H.A.; Balassa, J.J.; Tipton, H.J. Vanadium in plants. *J. Chronic Dis.* **1963**, *16*, 1047–1071. [CrossRef]
54. EU (2000) Position Paper: Ambient Air Pollution by AS, CD and NI Compounds. Working Group on Arsenic, Cadmium and Nickel Compounds. Available online: http://ec.europa.eu/environment/air/pdf/pp_as_cd_ni.pdf (accessed on 28 March 2017).
55. Bundes-Trinkwasserverordnung (BTrinkwV 2001) BGBl. I S. 2977: Verordnung Über die Qualität von Wasser für den Menschlichen Gebrauch [Regulation on the Quality of Water Intended for Human Consumption]. Available online: https://www.gesetze-im-internet.de/trinkwv_2001/BJNR095910001.html (accessed on 30 January 2021).
56. World Health Organization. *Guidelines for Drinking-Water Quality*, 4th ed.; World Health Organization: Geneva, Switzerland, 2011; pp. 1–632.
57. CLRTAP 2004. *Manual on Methodologies and Criteria for Modelling and Mapping Critical Loads and Levels and Air Pollution Effects, Risks and Trends*; UBA-Texte 52/2004: Dessau, Germany, 2004; pp. 1–266. Available online: https://www.umweltbundesamt.de/sites/default/files/medien/4038/dokumente/manual_complete_english.pdf (accessed on 28 March 2017).
58. Landesumweltamt Brandenburg-LUA Bbg. Vollzugshilfe zur Ermittlung erheblicher und irrelevanter Stoffeinträge in Natura 2000-Gebiete [Implementation Aid for the Identification of Significant and Irrelevant Substance Inputs into Natura 2000 areas]. Available online: https://mluk.brandenburg.de/sixcms/media.php/9/FFH-Vollzugshilfe-Stoffeintraege.pdf (accessed on 30 January 2021).
59. Madejón, P. *Thallium. Chapter 23 in Heavy Metals in Soils. Trace Metals and Metalloids in Soils and Their Bioavailability*; Alloway, B.J., Ed.; Springer: Dordrecht, The Netherlands; Heidelberg, Germany; New York, NY, USA; London, UK, 2013; pp. 543–547.
60. Scholl, G.; Metzger, F. Erhebungen über die Ti-Belastung von Nutzpflanzen auf kontaminierten Böden im Raum Lengerich [Surveys on the Ti contamination of crops on contaminated soil in the Lengerich area]. *J. Landwirtsch. Forsch.* **1981**, *38*, 216–223.

Article

First Results on Moss Biomonitoring of Trace Elements in the Central Part of Georgia, Caucasus

Omari Chaligava [1], Igor Nikolaev [2], Khetag Khetagurov [2], Yulia Lavrinenko [2], Anvar Bazaev [3], Marina Frontasyeva [1,*], Konstantin Vergel [1] and Dmitry Grozdov [1]

[1] Sector of Neutron Activation Analysis and Applied Research, Frank Laboratory of Neutron Physics, Joint Institute for Nuclear Research, 141980 Dubna, Russia; omar.chaligava@ens.tsu.edu.ge (O.C.); verkn@mail.ru (K.V.); dsgrozdov@rambler.ru (D.G.)

[2] Department of Anatomy, Physiology and Botany, Faculty of Chemistry, Biology and Biotechnology, The North Ossetian State University of K.L. Khetagurov, 362025 Vladikavkaz, Russia; bootany@yandex.ru (I.N.); zaz81@inbox.ru (K.K.); bionarium@nosu.ru (Y.L.)

[3] Department of Horticulture, Faculty of Agronomy, Gorsky State Agrarian University, 362040 Vladikavkaz, Russia; larix2@yandex.ru

* Correspondence: marina@nf.jinr.ru; Tel.: +7-903-260-6369

Abstract: The moss biomonitoring technique was used for assessment of air pollution in the central part of Georgia, Caucasus, in the framework of the UNECE ICP Vegetation. A total of 35 major and trace elements were determined by two complementary analytical techniques, epithermal neutron activation analysis (Na, Mg, Al, Cl, K, Ca, Sc, Ti, V, Cr, Mn, Fe, Co, Ni, Zn, Se, B, Rb, Sr, Zr, Mo, Sb, I, Cs, Ba, La, Ce, Nd, Sm, Eu, Tb, Yb, Hf, Ta, W, Th, and U) and atomic absorption spectrometry (Cu, Cd, and Pb) in the moss samples collected in 2019. Principal Component Analyses was applied to show the association between the elements in the study area. Four factors were determined, of which two are of geogenic origin (Factor 1 including Na, Al, Sc, Ti, V, Cr, Fe, Co, Ni, Th, and U and Factor 3 with As, Sb, and W), mixed geogenic–anthropogenic (Factor 2 with Cl, K, Zn, Se, Br, I, and Cu) and anthropogenic (Factor 4 comprising Ca, Cd, Pb, and Br). Geographic information system (GIS) technologies were used to construct distributions maps of factor scores over the investigated territory. Comparison of the median values with the analogous data of moss biomonitoring in countries with similar climatic conditions was carried out.

Keywords: moss biomonitoring; trace elements; atmospheric deposition; neutron activation analysis; atomic absorption spectrometry; multivariate statistics

Citation: Chaligava, O.; Nikolaev, I.; Khetagurov, K.; Lavrinenko, Y.; Bazaev, A.; Frontasyeva, M.; Vergel, K.; Grozdov, D. First Results on Moss Biomonitoring of Trace Elements in the Central Part of Georgia, Caucasus. *Atmosphere* **2021**, *12*, 317. https://doi.org/10.3390/atmos12030317

Academic Editor: Antoaneta Ene

Received: 30 December 2020
Accepted: 24 February 2021
Published: 28 February 2021

check for **updates**

1. Introduction

At present air pollution is recognized as the fifth largest threat to human health [1]. Air pollution and the associated problems are not confined by any geopolitical boundaries. The European Directives on air quality related to particulate matter (PM), heavy metals, and polycyclic aromatic hydrocarbons in ambient air [2,3], define target and limit values in the monitoring and further control of the pollutants. During the last several decades, biomonitoring surveys considering the use of an organism as a monitor of environmental pollution [4] have become a valuable complement to instrumental measurements. Widespread species that reliably reflect air pollution represent a simple and cost-effective alternative for instrumental measurements, thus enabling measurements with much higher spatial resolution. Mosses are recognized as good biomonitors of air pollution due to their specific morpho-physiological features: the lack of a root system, large surface area, and a high cation-exchange capacity of cell membranes, which represent their adaptations to nutrition from the air. Mosses are ubiquitous species and they have been extensively used in large-scale studies for biomonitoring of trans-boundary air pollution [5] known as passive moss biomonitoring [4]. The moss biomonitoring method, in combination with nuclear and related analytical techniques, has been regularly used for the last 25 years in Western European countries to study atmospheric deposition of heavy metals (HM). Over the past 15 years it has spread to Eastern Europe [5]. The first moss survey in Georgia was undertaken in 2014 [6] and the results included in the Report on the European Moss survey 2015–2016 [5] along with data obtained in the next surveys [7,8]. Rocks of different composition, age and stability are spread on the territory of Georgia. The high degree of the relief-dissection is due to strong tectonic movement and intense erosion processes in the Caucasus region. At certain locations of Georgia the depths of erosion-cuts exceeds 2000 m [9].

There is a need to investigate whether mosses sampled in this region can be used as biomonitors of atmospheric heavy metal deposition given the rather high contribution of mineral particles to the metal concentration in mosses. The present research was carried out by the Georgian and Russian teams aimed to cover white spots in the map of this territory of Caucasus.

2. Experimental

2.1. Study Area

The study area is located in Georgia between coordinates: 42°40′ N latitude and 43°17′ E longitude for the North, 41°22′ N latitude and 43°46′ E longitude for the South, 42°35′ N latitude and 43°13′ E longitude for the West, and 41°40′ N latitude and 44°41′ E longitude for the East. Elevation ranges from 651 to 2132 m a.s.l.

The South Caucasus region is highly prone to natural disasters, and its mountainous regions are particularly high risk areas. Natural phenomena common in the region include landslides and mudflows, floods, flash floods, droughts, avalanches, rainstorms, and earthquakes. The countries are located in a region of moderate to very high seismic activity and are therefore particularly prone to earthquakes, which can have devastating consequences for lives, buildings and infrastructure. This seismic activity can also trigger secondary events such as landslides, avalanches and flash floods in mountainous areas [10].

The mountainous regions of the South Caucasus have a wide range of climatic zones, from cold temperate alpine peaks to temperate, humid and arid landscapes.

The relief of Georgia is characterized by complex hypsometric and morphographic features: heavily dissected mountain slopes, deep erosive gorges, intermountain depressions, flat lowlands, plains, plateaus, and uplands. The most important landforms found in the territory of Georgia are erosive, volcanic, karst, gravitational, and old glacial landforms [11,12].

The climate in the high-mountains contributes to the formation of eternal snows and glaciers. Mountain meadow soils prevail in the highlands, and brown forest soils on the plains. Landscape and ecosystems of each sampling site differ considerably and depend on wind direction. Study area is located outside industrial zones; however, it may experience a long-range transport of pollutants due to resuspension of soil particles.

During the summer season, the main source of air pollution is traffic. It should be noted that as of 2018, 45.5% of vehicles were over 20 years old. Diesel fuel quality and requirements remain a particularly problematic issue in the country [13].

The main industrial activities taking place in the mountainous regions of Central Caucasus are related to the extraction and processing of natural resources. Mining activities alter the structure of the landscape, which can have severe consequences. Mining and processing activities often create toxic waste, which can have adverse impacts on the surrounding environment. In the Ambrolauri region, near village Uravi arsenic mining sites are situated. When the mining sites were abandoned in 1992, approximately 100,000 tons of wastes containing arsenic were left in surface areas. These sites are situated in the basins of the Rioni river, and there was an existing high risk of arsenic leakage [14,15].

2.2. Moss Sampling

Passive moss biomonitoring was performed in compliance with the guidelines of the Convention on Long-Range Transboundary Air Pollution (CLRTAP) and International Cooperative Program on Effects of Air Pollution on Natural Vegetation and Crops monitoring manual Moss Manual 2020 [16]. The following regions are represented: Racha-Lechkhumi and Kvemo Svaneti, Shida Kartli, Mtskheta-Mtianeti, Kvemo Kartli, and Samtskhe-Javakheti. Overall, thirty-five moss samples (*Hylocomium splendens* (Hedw.) Schimp. (4), *Hypnum cupressiforme* Hedw. (12)), *Pleurozium schreberi* (Brid.) Mitt (5), and *Abietinella abietina* (Hedw.) M. Fleisch) (14) were collected during summer 2019. (The number of samples of each type is given in brackets). Three first moss species are recommended for biomonitoring purposes in the Moss Manual-2020 [15], However in some sampling sited the only available species was *Abietinella abietina* (Hedw.) M. Fleisch) which was considered suitable for sampling due to the closeness of its morphological properties with the mosses listed in the Moss Manual. The sampling map is given in Figure 1. From a map Ecosystems of South Caucasus (Figure 2) one can obviously see the variety of ecosystems and climatic zones of the sampled areas.

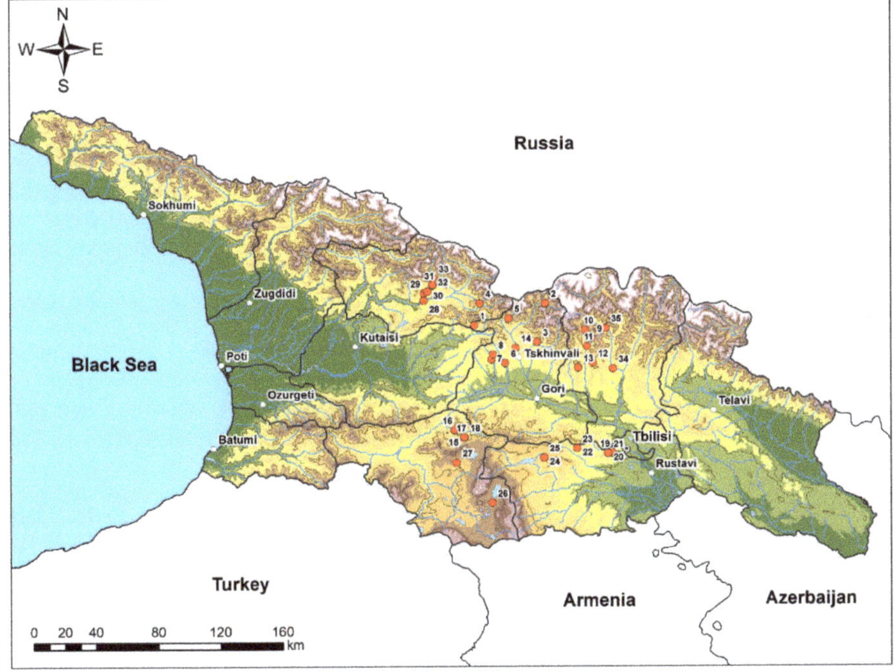

Figure 1. Sampling map.

Samples were collected at least 300 m from the main roads and settlements and at least 100 m away from the side roads, mainly from open areas to avoid the impact of higher vegetation. Longitude, latitude, and elevation were noted for every sampling location using the global positioning system.

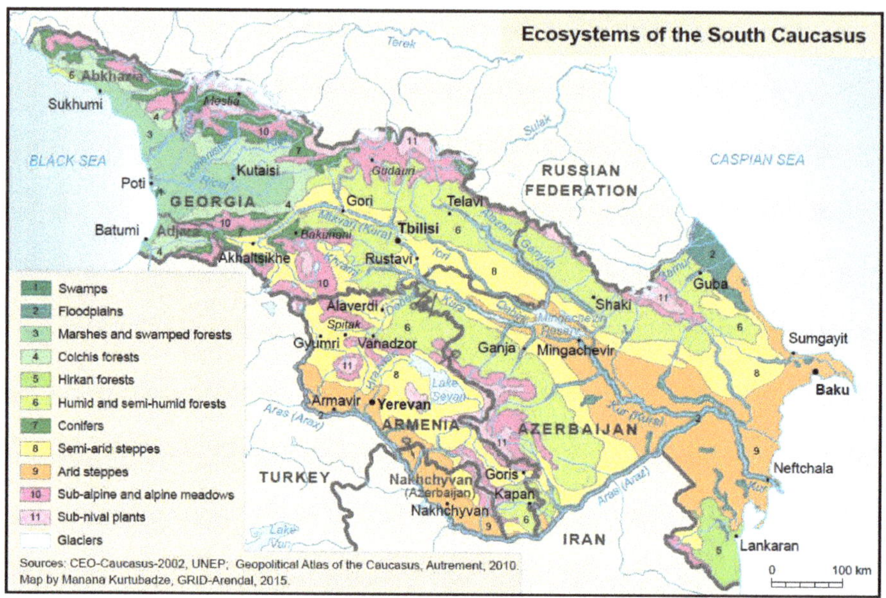

Figure 2. Ecosystems of South Caucasus [15].

Samples were collected at least 300 m from the main roads and settlements and at least 100 m away from the side roads, mainly from open areas to avoid the impact of higher vegetation. Longitude, latitude, and elevation were noted for every sampling location using the global positioning system.

For each sampling site, details (date of the sampling, weather condition, nearby vegetation, topography, and land use) were noted. Five to ten sub-samples were collected within an area of 50 m × 50 m and mixed in one composite sample. Samples were stored and transported in tightly closed paper bags. To prevent any contamination of the samples, sampling, and sample handling in the field and in the laboratory were performed using disposable polyethylene gloves (without talc) for each sample.

2.3. Sample Preparation and Elemental Analysis

Each sample was cleaned from extraneous materials in a chemical laboratory. Only green and green-brown shoots were taken and dried to a constant weight at 30–40 °C for 48 h. The elemental analysis of each sample was performed using instrumental neutron activation analysis (INAA) and atomic absorption spectrometry (AAS). The procedure of moss preparation for INAA and AAS is described in our previous study [8].

Moss samples were subjected to INAA at the neutron activation analysis facility REGATA of the IBR-2 reactor of the FLNP, JINR (Dubna, Russia). To determine elements with short lived isotopes (Mg, Al, Cl, Ca, Ti, V, Mn, I) samples were irradiated for 3 min and measured for 20 min. To determine elements with long lived isotopes (Na, K, Sc, Cr, Fe, Co, Ni, Zn, As, Se, Br, Rb, Sr, Zr, Mo, Sn, Sb, Cs, Ba, La, Ce, Nd, Sm, Eu, Tb, Yb, Hf, Ta, W, Au, Th and U) samples were irradiated for 3 days, re-packed, and measured twice using HPGe detectors after 4 and 20 days of decay, respectively. The calculation of element concentrations was performed using software developed at FLNP JINR [17].

The AAS was used to determine amounts of Cu, Cd, and Pb in the moss samples using the iCE 3300 AAS atomic absorption spectrometer with electrothermal (graphite furnace) atomization (Thermo Fisher Scientific, Waltham, MA, USA).

The calibration solutions were prepared from a 1 g/L stock solution (AAS standard solution; Merck, DE).

2.4. Quality Control of ENAA and AAS

In order to evaluate the precision and accuracy of the results, the certified reference materials and standards were used, namely NIST SRM 1575a—Trace Elements in Pine Needles, NIST SRM 1547—Peach Leaves, NIST SRM 1633b—Constituent Elements in Coal Fly Ash, NIST SRM 1632c—Trace Elements in Coal (Bituminous), IRMM SRM 667—Estuarine Sediment, NIST SRM 2711—Montana Soil, NIST SRM 2710—Montana Soil.

Table 1 shows the differences between certified and calculated values of concentrations, where "SRM" were used as standards for calculations of concentrations for SRMs in the column "Sample". Most differences between certified and obtained values are lower than 2 σ. There are no such data for elements Mo, Sn, W, and Au because their certified values are in the irradiated SRM only.

Table 1. Epithermal Neutron Activation Analysis (ENAA): obtained and certified values of reference materials, mg/kg.

SRM	Sample	Element	Obtained	Certified	SRM	Sample	Element	Obtained	Certified
2709a	FFA1	Na	21,647 ± 1602	21,900 ± 811	FFA1	2709a	Rb	100.4 ± 16.6	99.00 ± 2.97
1547	1575a	Mg	1086.5 ± 59.8	1060 ± 170	FFA1	2709a	Sr	239.8 ± 21.1	239.00 ± 5.98
1632c	1633c	Al	133,841 ± 4016	132,800 ± 6109	2709a	1632c	Zr	18.31 ± 4.86	16.0 ± 4.8
1549	1632c	Cl	1120 ± 77	1139 ± 41	2709a	FFA1	Sb	17.92 ± 3.79	17.6 ± 2.5
2709a	FFA1	K	22,498 ± 1373	22,000 ± 6600	1549	1547	I	0.394 ± 0.099	0.30 ± 0.09
1633c	2709a	Ca	19,711 ± 1814	19,100 ± 898	2709a	667	Cs	7.9 ± 0.3	7.80 ± 0.71
2709a	667	Sc	13.82 ± 0.35	13.7 ± 0.7	2709a	1632c	Ba	41.83 ± 6.03	41.1 ± 1.61
1633c	2710a	Ti	2928 ± 205	3110 ± 72	667	2709a	La	21.26 ± 1.13	21.7 ± 0.4
1633c	1632c	V	22.79 ± 1.03	23.72 ± 0.53	667	2709a	Ce	42.26 ± 2.79	42.00 ± 1.01
FFA1	667	Cr	181.6 ± 11.9	178 ± 16	667	FFA1	Nd	51.44 ± 6.49	56.8 ± 3.7
1575a	1547	Mn	97.83 ± 5.97	98 ± 3	667	1632c	Sm	1.021 ± 0.092	1.078 ± 0.029
FFA1	2709a	Fe	34,306 ± 1784	33,600 ± 706	667	2709a	Eu	0.833 ± 0.067	0.83 ± 0.02
2709a	FFA1	Co	39.87 ± 1.36	39.8 ± 1.7	667	FFA1	Tb	1.285 ± 0.047	1.38 ± 0.14
2709a	1632c	Ni	9.56 ± 0.76	9.32 ± 0.52	FFA1	667	Yb	2.49 ± 0.23	2.200 ± 0.091
2709a	667	Zn	177.11 ± 9.04	175 ± 13	FFA1	1632c	Hf	0.508 ± 0.047	0.585 ± 0.011
FFA1	1632c	As	6.07 ± 0.36	6.18 ± 0.28	667	FFA1	Ta	1.774 ± 0.055	2.11 ± 0.17
1632c	2709a	Se	1.02 ± 0.18	1.5 ± 0.45	FFA1	2709a	Th	10.95 ± 0.43	10.9 ± 0.2
667	1632c	Br	21.08 ± 0.85	18.7 ± 0.4	2709a	1632c	U	0.51 ± 0.03	0.52 ± 0.02

A comparison of heavy metal concentrations obtained using the AAS with the standard values are presented in Table 2. The difference between the certified and measured elements contents of the certified material varied between 1% and 5%.

Table 2. Comparison of the atomic absorption spectrometry (AAS)-obtained heavy metal concentrations with the standard values, mg/kg.

Element	Certified	Obtained
Cd	0.54 ± 0.02	0.54 ± 0.01
Cu	5.0 ± 0.10	4.6 ± 0.3
Pb	0.20 ± 0.06	0.18 ± 0.01

2.5. Data Analysis Using PCA

Principle Component Analysis (PCA) is a special case of factor analysis, which transforms the original set of intercorrelated variables into a set of uncorrelated variables that are linear combinations of the original variables. The first principal component is the linear combination of the variables that accounts for a maximum of the total variability of the data set. The second principal component explains a maximum of the variability not accounted for by the first component, and so on. The objective is to find a minimum number of principal components that explain most of the variance in the data set. The principal components are statistically independent and, typically, the first few components explain almost all the variability of the whole data set. The minor principal components, which explain only a minor part of the data, can be eliminated, thus simplifying the analysis. Further, these minor components contain most of the random error, so eliminating them tends to remove extraneous variability from the analysis. A wide range in concentrations makes normalization of the data necessary if all the elements are to be given equal weight in the analysis. The values used in the PCA are made dimensionless by this transformation [18]).

2.6. Construction of GIS Maps

The ArcGis 10.6 software (Esri, Redlands, California, USA) was used to build distribution maps of factor scores over the study area. We are using the OpenLayers library and a few backgrounds like "Oceans", "Gray", "World", "OSM", etc., to generate maps.

3. Results and Discussion

A summary of the results from the 2019 moss sampling over the study area is presented in Table 3 along with similar data obtained in previous surveys in Georgia in 2014–2017 [8], North Macedonia [19], Bulgaria [20], and pristine country Norway [21]. Data from North Macedonia and Bulgaria were obtained by INAA in Dubna, at the IBR-2 reactor of FLNP JINR using the same hard- and software, whereas Norwegian data is a result of ICP-MS. The Table 3 contains the medians and the lower and upper concentration quartiles of all components. Variability of elemental concentrations is reflected by the total range, which often spans approximately two to three orders of magnitude. Direct comparison of the medians does not show great difference in the elemental concentrations for Georgia and the Balkan countries, whereas maximal values of such element as As and Mo, both in 2014–2017 and 2019, exceed those for North Macedonia and Bulgaria, and it is five times higher than the maximum in Norway. This phenomenon is easily explained by mining and processing of arsenic and the presence of polymetallic ores abundant in the Caucasian Mountains. To demonstrate special behavior of As, the Summary Results for arsenic, iron, zinc, and nickel are presented in Figure 3 from which a strong local As contamination is evident, whereas Fe shows normal distribution, and the others are close to normal. In comparison with Norway, a country with fewer anthropogenic influences, Georgia has higher median values for the elemental content in mosses for almost all air pollution elements (As, Cd, Co, Cr, Cu, Hg, Ni, and Pb) [21].

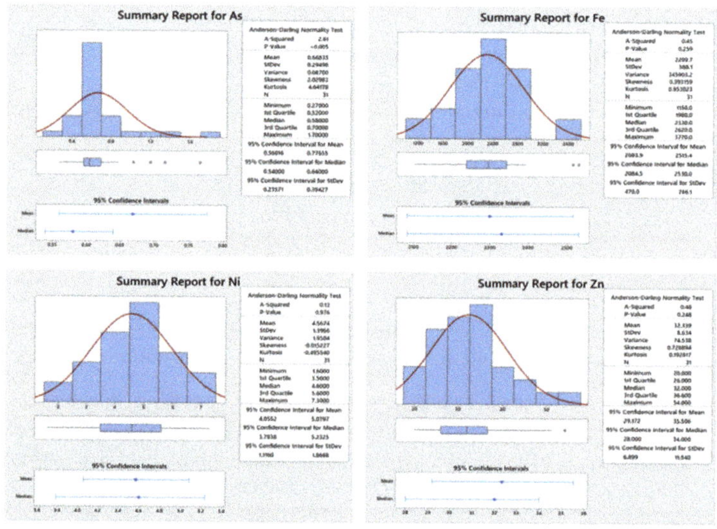

Figure 3. Summary Report for As and some selected elements created byMinitab® 19.

Table 3. Comparison of the results obtained in present study with the countries allocated in relatively the same geographical belt. Norway chosen as a pristine area. Concentration is in mg/kg.

Element	Georgia, 2019 Present Study		Georgia, 2014–2017 [8]		North Macedonia, 2015 [19]		Bulgaria, 2015/16 [20]		Norway, 2015 [21]	
n—number of samples	n = 35		n = 120		n = 72		n = 115		n = 229	
Element	Median	Range	Median	Range	Median	Range	Median	Range	Median	Range
Na	482	169–1350	581	101–3000	190	140–380	225	79–1560	210	60–800
Mg	2640	1910–4420	3060	1220–11600	1900	1200–3800	2080	514–8550	1350	470–3280
Al	2770	1610–9680	4295	759–24,500	2100	750–7400	2310	569–10,900	460	100–3050
Cl	132	56–635	185	57.3–1080	ND	ND	78.8	16.6–861	ND	ND
K	5930	3970–8860	5935	2030–15,000	6000	3100–14,000	5670	3250–14,200	3560	1770–6400
Ca	8400	5490–12,400	8255	4620–17,100	6900	3500–13,000	6630	606–14,200	3030	1820–7230
Sc	0.8	0.36–2.1	1.11	0.17–6.58	ND	ND	0.41	0.10–3.13	0.09	0.02–1.4
Ti	238	129–596	349.5	68.6–2100	ND	ND	143	46.4–764	24	6–152
V	6.5	3.85–16.2	9.4	1.71–54	3.3	0.47–11	3.89	1.3–22.7	1.2	0.3–14
Cr	6.6	3.14–15.5	7.75	14,337	5.7	11,536.00	2.73	0.219–25	0.7	0.2–17
Mn	142	64–377	141	230,306	160	33–510	180	39–551	400	40–1660
Fe	2410	1150–6110	2725	404–14,100	1700	510–4600	1190	376–7240	310	78–8125
Co	1	0.36–2.6	1.43	0.23–8.12	0.6	0.16–2	0.59	0.197–3.29	0.2	0.06–23
Ni	4.9	1.6–10.8	5.56	1.92–24.2	3.5	0.68–63	2.1	0.45–13.5	1.1	0.4–550
Cu *	8.93	6.58–13.7	5.54	0.13–143	4.6	3.0–8.3	7.36	3.2–46.88	4.2	1.8–370
Zn	32	20–54	28.85	7.15–75.2	30	.66	28	9–101	31	8–409
As	0.58	0.27–27.6	1.05	0.18–83.3	0.54	0.13–1.4	0.45	0.20–3.57	0.13	0.04–4.72
Se	0.18	0.09–0.37	0.23	0.068–0.65	ND	ND	0.2	0.008–0.67	0.3	0.009–2
Br	4.4	1.84–8.1	6.31	2.33–25.2	ND	ND	2.8	1.2–9.4	ND	ND
Rb	7.2	4.1–16.4	10.55	2.92–34.2	5.3	02.02.2028	7.38	2.24–50.7	12.4	1.4–81
Sr	37.6	24–68	43.85	17.2–157	25	6.5-220	25	11.3–122	13.6	3.8–60
Zr	9.4	3.5–25	10.55	1.19–67.9	ND	ND	ND	ND	ND	ND
Mo	0.57	0.37–6.3	0.35	0.14–2.1	0.17	0.08–0.51	ND	ND	ND	ND
Element	Median	Range	Median	Range	Median	Range	Median	Range	Median	Range
Cd *	0.12	0.058–0.35	5.57	0.01–0.58	0.23	0.018–0.88	0.1	0.02–1.56	0.08	0.02–1.33
Sb	0.13	0.07–0.72	0.15	0.049–1.36	ND	ND	0.11	0.04–0.51	0.07	0.007–0.38
I	2.04	1.14–4.35	0.16	0.58–11.8	ND	ND	1.28	0.48–2.99	ND	ND
Cs	0.263	0.16–0.97	2.48	0.036–2.67	ND	ND	0.207	0.0716–1.8	0.16	0.02–1.63
Ba	37	17–100	0.42	4.98–365	42	9.7–180	46	14.2–309	25	5.3–130
La	1.57	0.82–6.2	51.3	0.34–12	ND	ND	1.35	0.39–22.6	0.32	0.07–3.5
Ce	2.97	1.38–9.2	2.15	0.31–21.7	ND	ND	2.4	0.5–29.2	0.61	0.10–4.78
Nd	1.73	0.87–7.6	3.75	0.45–10.7	ND	ND	1.3	0.2–24.1	0.23	0.01–2.24
Sm	0.26	0.12–1.17	2.04	0.031–2.7	ND	ND	ND	ND	0.05	0.004–0.38
Eu	0.054	0.02–0.22	0.34	0.023–0.52	ND	ND	0.07	0.009–0.92	0.04	0.01–0.19
Tb	0.04	0.02–0.15	0.1	0.011–0.31	ND	ND	0.03	0.005–0.42	0.01	<0.001–0.09
Yb	0.15	0.054–0.51	0.05	0.022–0.8	ND	ND	0.1	0.03–1.08	0.003	<0.001–0.016
Hf	0.24	0.11–0.76	0.15	0.041–1.81	ND	ND	0.16	0.04–1.44	ND	ND
Ta	0.046	0.02–0.13	0.27	0.0069–0.28	ND	ND	0.04	0.009–0.28	ND	ND
W	0.14	0.04–0.41	0.06	0.026–0.67	ND	ND	0.1	0.02–1.44	ND	ND
Pb *	4.33	2.34–8.21	0.11	0.18–9.1	4.9	2.2–14	10.7	3.72–102.8	0.05	0.001–0.4
Th	0.45	0.18–1.57	0.51	0.063–2.9	ND	ND	0.39	0.09–2.8	0.03	0.007–1.5
U	0.12	0.06–0.34	0.16	0.021–1.25	ND	ND	0.12	0.03–3.2	0.006	0.002–0.08

* Elements determined by AAS marked with asterisks.

It is also clearly confirmed by principal component analysis (PCA) used to classify the elements with respect to contribution sources.

PCA was carried out by using the Statistical Package STATISTICA 13.0 Results of factor analysis are presented in Table 4. Communality values close to 1 suggest that the extracted factors explain much of the variance of the individual variable.

Table 4. Rotated factor loadings for the Central Georgia data set (36 samples). Varimax normalized. Extraction: Principal components (Marked loadings are > 0.6).

Variable	Factor 1	Factor 2	Factor 3	Factor 4	Communality
Na	0.72	0.21	0.53	0.21	0.98
Al	0.82	−0.16	−0.15	−0.43	1.00
Cl	−0.10	0.83	0.04	0.19	0.92
K	−0.08	0.87	0.18	−0.06	0.81
Ca	0.27	0.03	-0.05	−0.80	0.80
Sc	0.93	0.07	0.19	0.14	1.00
Ti	0.93	−0.07	0.21	−0.12	0.98
V	0.86	−0.07	−0.07	−0.30	0.99
Cr	0.90	0.14	0.18	0.19	0.99
Fe	0.94	0.05	0.20	0.11	1.00
Co	0.69	0.01	0.38	0.28	0.94
Ni	0.67	0.34	0.32	0.40	0.97
Zn	0.10	0.78	0.20	−0.12	0.83
As	0.30	0.07	0.89	−0.12	0.97
Se	0.07	0.69	−0.09	0.54	0.92
Br	0.19	0.65	0.00	0.65	0.94
Sb	0.34	0.24	0.85	0.00	0.98
I	0.18	0.74	−0.08	0.54	0.96
Th	0.90	−0.05	0.29	0.06	0.89
U	0.85	0.05	0.41	0.15	1.00
Cd	0.27	0.49	0.10	0.64	1.00
Pb	0.19	0.26	0.12	0.71	0.91
Cu	−0.07	0.80	0.06	0.27	0.92
W	0.26	0.03	0.72	0.32	0.84
Expl.Var	8.39	4.75	3.10	3.49	19.72
Prp.Totl	0.35	0.20	0.13	0.15	0.82

The data set analyzed includes results for 24 trace elements and major components. The PCA indicates four factors, which explain 82% of the total variance.

To visualize the results obtained, the graph on Factor Loadings was built (see Figure 4).

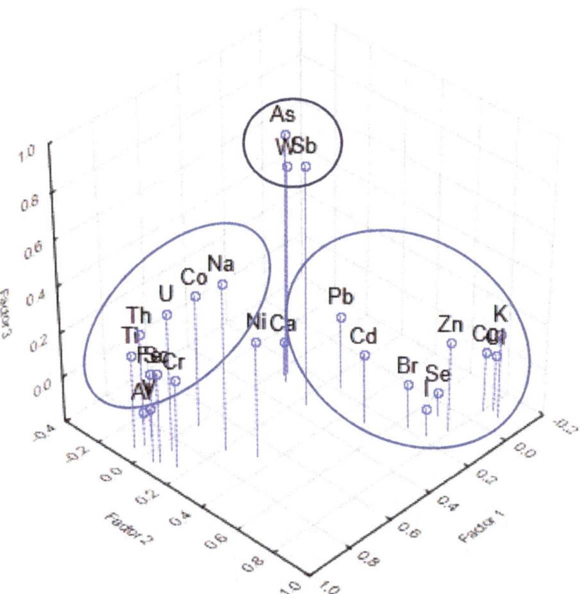

Figure 4. Factor Loadings, Factor 1 vs. Factor 2 vs. Factor 3. Rotation: Varimax normalized. Extraction: Principal components.

The results of factor scores are presented in the form of distribution geographic information system (GIS) maps. See Figure 4.

Factor 1 is loaded with Na, Al, Sc, Ti, V, Cr, Fe, Co, Ni, Th, and U, and represents mainly a combination of light and heavy crust component elements in the form of soil dust. It has almost 35% of the total variability and is the strongest factor (Figure 5). The contents of these elements in the moss samples are significantly influenced by the mineral particles that are carried into the atmosphere by winds, and their spatial distribution mainly depends on urban activities that are not related to industrial activities. High contents of elements of this geochemical association have been found in samples taken from the sampling points 28–29 (Racha-Lechkhumi and Kvemo Svaneti region, Ambrolauri municipality); 4 (Racha-Lechkhumi and Kvemo Svaneti region, Oni municipality); 10 (Mtskheta-Mtianeti region, Akhalgori municipality); 13 (Mtskheta-Mtianeti region, Akhalgori municipality); 22–23 (Kvemo Kartli region, Tetritskaro municipality); and 26 (Samtskhe-Javakheti region, Ninotsminda municipality).

Factor 2 contains Cl, K, Zn, Se, Br, I, and Cu and represents a combination of two sub-factors, a marine one: halogens Cl, Br, I and Se [22], and the second one possibly is due of some local agricultural activity. Zinc, potassium and copper are essential elements for several biochemical processes in plants [23]. The concentrations of heavy metals such as zinc and copper in the environment are currently increasing, due mainly to human activities. Copper is still used for protecting purposes in agriculture: it prevents and cure diseases, which can have adverse effects on crop yields and quality. Factor 3 includes As (0.89), Sb (0.85), and W (0.72) which are characteristic for ores used for arsenic extraction. In particular, in the village of Uravi (Ambrolaur region, Western Georgia) a mining and chemical factory functioned during the Soviet era. Arsenic has been mined and processed there for almost 60 years.

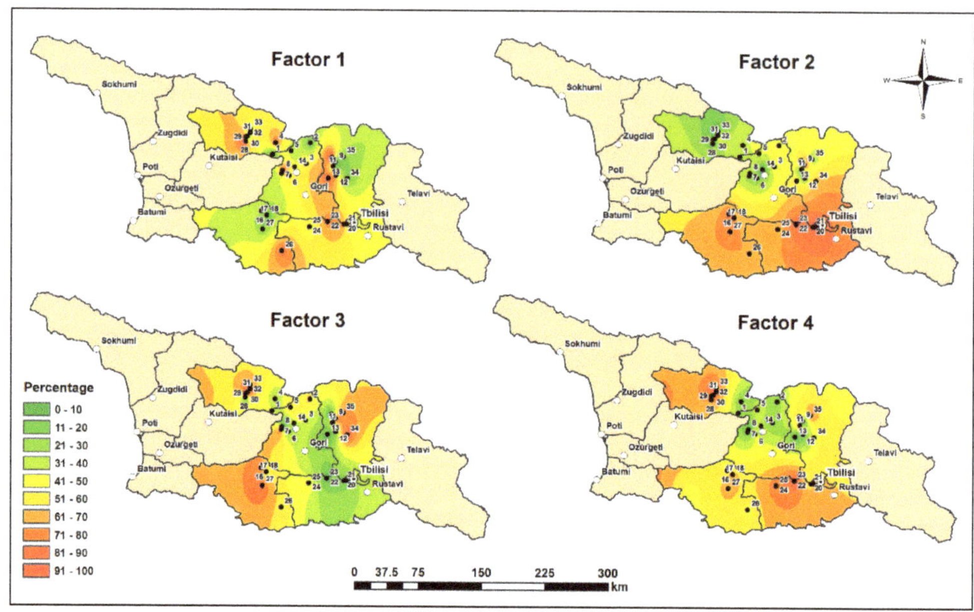

Figure 5. Factor Scores.

Factor 4 is represented by Ca (0,80) Cd (0.64), Pb (0.64), and Br (0.65) of local anthropogenic origin due to closeness to urban areas. Lead and cadmium enter the environment in the form of impurities in fertilizers, halides and oxides of these metals, as well as bromides which are contained in the exhaust gases of cars, as part of the waste generated during the extraction and processing of used batteries [24]. The highest contents of these elements are found in the moss samples collected from the sampling points 28–33 (Racha-Lechkhumi and Kvemo Svaneti region, Ambrolauri municipality); 17–18 (Samtskhe-Javakheti region, Borjomi municipality); 27 (Samtskhe-Javakheti region, Akhalkalaki municipality); 22–23 (Kvemo Kartli region, Tetritskaro municipality); 24–25 (Kvemo Kartli region, Tsalka municipality); 9 (Mtskheta-Mtianeti region, Akhalgori municipality); and 35 (Mtskheta-Mtianeti region, Dusheti municipality).

4. Conclusions

For the first time atmospheric deposition of trace elements using moss biomonitoring technique was studied in Central Georgia in 2019. By the comparison of the obtained values for a broad set of elements (Al, As, Ba, Ca, Cd, Co, Cr, Cu, Fe, Hg, K, Li, Mg, Mn, Mo, Ni, Na, Pb, Rb, Sr, V, Zn, Th, and U) with the data from previous surveys in other parts of Georgia and in the countries of the similar climatic conditions (North Macedonia and Bulgaria) it was shown that air pollution in Central Georgia does not exceed mean values for these European countries, whereas data for potentially toxic elements such as As, Cd, Co, Cr, Cu, Ni, Pb, and Zn exceed the ones in Norway used as an example of a pristine country of Europe. Of the four factors, determined by PCA one factor (F4) is purely anthropogenic (Ca, Cd, Pb, and Br) and it is explained by the relevant high factor scores in the urban areas where they may come from fertilizers, halides, and oxides as well as bromides of these metals, which are contained in the exhaust gases of cars, as part of the waste generated during the extraction and processing of used batteries, etc. High As and W loadings if factor 3 are explained by intense mining activity for more than 60 years of As extraction from ores rich in this element and accompanying elements such as antimony and tungsten. A strong marine component (Cl, Br, I, and Se) in factor 2 is provided by the location of Georgia between two seas—the Black and the Caspian ones. In this factor

2 elements of marine component are mixed with Zn, K, and Cu due to most probably agricultural activity. Factor 1 represents mainly a combination of light and heavy crust component elements in the form of soil dust.

Author Contributions: Conceptualization, O.C., I.N., K.K., Y.L. and M.F.; Formal analysis, O.C.; Investigation, I.N., K.K., Y.L., A.B., K.V. and D.G.; Methodology, M.F.; Software, O.C.; Writing—original draft, M.F.; Writing—review & editing, I.N. and K.K. All authors have read and agreed to the published version of the manuscript.

Funding: This research was supported by the grant of the Plenipotentiary of Georgia at JINR in 2020 (Order #37 of 22.01.2020).

Data Availability Statement: Not applicable.

Acknowledgments: This research was supported by the grant of the Plenipotentiary of Georgia at JINR in 2020 (Order #37 of 22.01.2020). The authors express their gratitude to Richard Hoover for editing the English.

Conflicts of Interest: The authors declare no conflict of interest.

References

1. Cohen, A.; Brauer, M.; Burnett, R.; Anderson, H.R.; Frostad, J.; Estep, K.; Balakrishnan, K.; Brunekreef, B.; Dandona, L.; Dandona, R.; et al. Estimates and 25-year trends of the global burden of disease attributable to ambient air pollution: An analysis of data from the Global Burden of Diseases Study 2015. *Lancet* **2017**, *389*, 1907–1918. [CrossRef]
2. Directive 2004/107/EC of the European parliament and of the council of 15 December 2004 relating to arsenic, cadmium, mercury, nickel and polycyclic aromatic hydrocarbons in ambient air. *Off. J. Eur. Union* **2005**, *23*, 1–16.
3. Directive2008/50/EC of the European parliament and of the council of 21 May 2008 on ambient air and cleaner air for Europe. *Off. J. Eur. Union* **2008**, *152*, 1–44.
4. Markert, B.A.; Breure, A.M.; Zechmeister, H.G. Definitions, strategies and principles for bioindications/biomonitoring of the environment. In *Bioindicators & Biomonitors*; Markert, B.A., Breure, A.M., Zechmeister, H.G., Eds.; Elsevier Science Ltd.: Amsterdam, The Netherlands, 2003; pp. 3–41.
5. Frontasyeva, M.; Harmens, H.; Uzhinskiy, A.; Chaligava, O.; Participants of the Moss Survey. *Mosses as Biomonitors of Air Pollution: 2015/2016 Survey on Heavy Metals, Nitrogen and POPs in Europe and beyond*; Report of the ICP Vegetation Moss Survey Coordination Centre; Joint Institute for Nuclear Research: Dubna, Russian, 2020; p. 136. ISBN 978-5-9530-0508-1.
6. Shetekauri, L.S.; Shetekauri, T.; Kvlividze, A.; Chaligava, O.; Kalabegeshvili, T.; Kirkesali, E.I.; Frontasyeva, M.V.; Chepurchenko, O.E. Preliminary results of atmospheric deposition of major and trace elements in the Greater and Lesser Caucasus Mountains studied by the moss technique and neutron activation analysis. *Ann. Bot.* **2015**, 89–95. [CrossRef]
7. Shetekauri, S.; Chaligava, O.; Shetekauri, T.; Kvlividze, A.; Kalabegishvili, T.; Kirkesali, E.; Frontasyeva, M.V.; Chepurchenko, O.E.; Tselmovich, V.A. Biomonitoring air pollution with moss in Georgia. *Polish J. Environ. Stud.* **2018**, *27*, 2259–2266. [CrossRef]
8. Chaligava, O.; Shetekauri, S.; Badawy, W.M.; Frontasyeva, M.V.; Zinicovscaia, I.; Shetekauri, T.; Kvlividze, A.; Vergel, K.; Yushin, N. Trace elements in atmospheric deposition studied by moss biomonitoring using neutron activation analysis and atomic absorption spectrometry in Georgia: Characteristics and impacts. *Arch. Environ. Contam. Toxicol.* **2020**. [CrossRef]
9. Van Westen, C.J.; Straatsma, M.; Turdukulov, U.; Feringa, W.F.; Sijmons, K.; Bakhtadze, K.; Janelidze, T.; Kheladze, N. *Atlas of Natural Hazards and Risk in Georgia*; CENN Caucasus Environmental NGO Network: Tbilisi, Georgia, 2012; ISBN 978-9941-0-4310-9. (In Georgian)
10. UNISDR (United Nations Office for Disaster Risk Reduction). *Central Asia and Caucasus Disaster Risk Management Initiative (CAC DRMI). Risk Assessment for Central Asia and Caucasus: Desk Study Review*; The United Nations Office for Disaster Risk Reduction: Geneva, Switzerland, 2009. (In Russian)
11. Gulisashvili, V.Z. *Natural Zones and Natural-Historical Regions of the Caucasus*; Nauka: Moscow, Russia, 1964; 327p. (In Russian)
12. Melikishvili, G.A.; Lordkipanidze, O.D.; Academy of Sciences of the Georgian SSR. Essays on the history of Georgia. In *Javakhishvili Institute of History, Archeology and Ethnography. Date of Admission to EC 27.02.2002*; Metsniereba: Tbilisi, Georgia, 1989; Volume 8. (In Russian)
13. Special Report—"Right to Clean Air (Air Quality in Georgia)". Public Defender (Ombudsman) of Georgia. Available online: https://www.ombudsman.ge/geo/spetsialuri-angarishebi/190627034307spetsialuri-angarishi-ufleba-sufta-haerze-atmosferuli-haeris-khariskhi-sakartveloshi (accessed on 30 December 2020).
14. Gurguliani, I. Arsenic Contamination in Georgia, Environment Related Security Risks from old Soviet Legacy. In Proceedings of the 22nd OSCE Economic and Environmental Forum "Responding to Environmental Challenges with a View to Promoting Cooperation and Security in the OSCE Area", Vienna, Austria, 27–28 January 2014; Available online: https://www.osce.org/files/f/documents/a/a/110702.pdf (accessed on 30 December 2020). (In Russian).
15. Alfthan, B. *Outlook on Climate Change Adaptation in the South Caucasus Mountains—2016*. 2016. Available online: https://www.weadapt.org/sites/weadapt.org/files/2017/june/caucasus_smd.pdf (accessed on 30 December 2020). (In Russian).

16. Moss Manual "Heavy Metals, Nitrogen and POPs in European Mosses: 2020 Survey". Available online: https://icpvegetation.ceh.ac.uk/sites/default/files/ICP%20Vegetation%20moss%20monitoring%20manual%202020.pdf (accessed on 30 December 2020).

17. Pavlov, S.S.; Dmitriev, A.Y.; Chepurchenko, I.A.; Frontasyeva, M.V. Automation system for measurement of gamma-ray spectra of induced activity for neutron activation analysis at the reactor IBR-2 of Frank Laboratory of Neutron Physics at the Joint Institute for Nuclear Research. *Phys. Elem. Part Nucl.* **2014**, *11*, 737–742. Available online: https://link.springer.com/article/10.1134/S1547477114060107 (accessed on 30 December 2020). [CrossRef]

18. Schaug, J.; Rambaek, J.P.; Steinnes, E.; Henry, R. Multivariate Analysis of Trace Element Data from Moss Samples Used to Monitor Atmospheric Deposition. *Atmos. Environ.* **1990**, *24A*, 2625–2631. [CrossRef]

19. Stafilov, T.; Šajn, R.; Barandovski, L.; Andonovska, K.B.; Malinovska, S. Moss biomonitoring of atmospheric deposition study of minor and trace elements in Macedonia. *Air Qual. Atmos. Health* **2018**, *11*, 137–152. [CrossRef]

20. Hristozova, G.; Marinova, S.; Svozilík, V.; Nekhoroshkov, P.; Frontasyeva, M.V. Biomonitoring of elemental atmospheric deposition: Spatial distributions in the 2015/2016 moss survey in Bulgaria. *J. Radioanal. Nucl. Chem.* **2019**. [CrossRef]

21. Steinnes, E.; Uggerud, H.T.; Pfaffhuber, K.A.; Berg, T. *Atmospheric Deposition of Heavy Metals in Norway*; National Moss Survey 2015; NILU—Norwegian Institute for Air Research: Kjeller, Norway, 2016; p. 55.

22. Frontasyeva, M.V.; Steinnes, E. Marine gradients of halogens in moss studied by epithermal neutron activation analysis. *J. Radioanal. Nucl. Chem.* **2004**, *261*, 101–106. [CrossRef]

23. Kabata-Pendias, A. *Trace Elements in Soils And Plants*, 4th ed.; CRC Press/Taylor & Francis Group: Boca Raton, FL, USA, 2010; ISBN 9780429192036,. [CrossRef]

24. Nriagu, J.O. A global assessment of natural sources of atmospheric trace metals. *Nature* **1989**, *338*, 47–49. [CrossRef]

Article

Estimating Background Values of Potentially Toxic Elements Accumulated in Moss: A Case Study from Switzerland

Stefano Loppi [1,*] **, Zaida Kosonen** [2] **and Mario Meier** [2]

[1] Department of Life Sciences, University of Siena, 53100 Siena, Italy
[2] Research Center for Environmental Monitoring, 8640 Rapperswil, Switzerland; zaida.ehrenmann@fub-ag.ch (Z.K.); mario.meier@fub-ag.ch (M.M.)
* Correspondence: stefano.loppi@unisi.it

Abstract: Although the use of moss as biomonitor of air pollution is relatively simple, the interpretation of the data needs reference values. Background values for Cd, Cu, Pb, and Zn accumulated in moss samples from Switzerland, collected every five years from 1995 to 2015 in the framework of the European Moss Survey, were statistically estimated. These background values can be used as reference for the assessment of spatial and temporal trends, to be expressed in terms of bioaccumulation ratios with actual values. The use of annual background values is of great importance to identify spatial trends, while period-wide background values identify temporal trends. The latter are consistent with those reported in other comprehensive similar biomonitoring studies in Europe and are required to be updated in time, possibly every five years. The use of cutoff values to be used as benchmark for bioaccumulation ratios is invaluable in having a scale for assessing ecological quality.

Keywords: biomonitors; PTE; temporal change; atmospheric pollution; deposition; heavy metal

Citation: Loppi, S.; Kosonen, Z.; Meier, M. Estimating Background Values of Potentially Toxic Elements Accumulated in Moss: A Case Study from Switzerland. *Atmosphere* **2021**, *12*, 177. https://doi.org/10.3390/atmos12020177

Academic Editor: Antoaneta Ene
Received: 29 December 2020
Accepted: 26 January 2021
Published: 29 January 2021

1. Introduction

The term "heavy metal", although widely used in the literature, is now deprecated and it has been suggested to replace it with the term of potentially toxic element (PTE), especially in the case of environmental studies [1]. Having a wide variety of emission sources such as motor vehicles, heating systems, industrial plants, etc., PTEs are an important component of air pollution and have a great epidemiological concern because of their persistence in the environment and the negative effects on human health [2].

Since about 80% of the EU population lives in urban areas, urban air pollution affects the quality of life of most citizens. Consequently, urban air pollution has largely been investigated (see, e.g., [3]) and many urban areas have an air quality monitoring network to monitor whether the set environmental standards are met or not. However, studies on atmospheric deposition at remote areas to evaluate the impact of PTEs on ecosystems are less common [4,5] and an evaluation on background pollution in pristine areas is only very rarely accessed. This is often due to economic constraints related to the establishment and maintenance of sophisticated and costly equipment. In such cases, the use of living organisms may be very useful to complement the data obtained by physico-chemical measurements. Since any change taking place in the environment has a significant effect on the biota, biological monitoring (biomonitoring) is a very effective early warning system to detect environmental changes [6].

Carpet-forming moss species are among the most valuable biomonitors of atmospheric pollution as they are highly dependent on wet and dry atmospheric deposition for nutrients and lack a waxy cuticle and stomata, allowing the absorbance of contaminants over the whole moss surface [7]. Additionally, mosses can accumulate persistent pollutants, such as PTEs and are used to measure the amounts of pollutants in the ecosystem that are biologically available [8]. The abundance of these moss species in Scandinavia gave rise to the beginning of the moss monitoring survey making it possible to monitor PTE pollution on a multicountry-wide scale [9]. Thanks to the standardized method and the relatively low costs as well as the ease of collection of samples, this well-established technique of moss monitoring has been adopted also in other European countries coordinated by an International Cooperative Program on Effects of Air Pollution on Natural Vegetation and Crops (ICP Vegetation) under the United Nations Economic Commission for Europe (UNECE) Geneva Air Convention. Through this program, mosses have been used to evaluate atmospheric deposition of PTEs at several European countries every five years since 1990 [10]. This enabled not only the comparison of spatial patterns of PTE deposition, but also the detection of temporal trends. Presently, also other pollutants have been added to the survey such as nitrogen, persistent organic pollutants (POPs), and, more recently, also microplastics. Although outside of Europe mosses have been sometimes used as biomonitors of air pollution, the wide use of the moss monitoring technique is mainly restricted to Europe, where ca. 80% of the studies have been performed [11].

Switzerland has been participating in the moss monitoring project since 1990 and the Swiss outcomes showed a general decline in element concentrations in time [12], likely determined by the closure of several small industries, mostly metallurgic, as well as improved abatement technology of waste incinerators, the ban of leaded fuel, and the wide use of catalytic converters in cars. This hypothesis is corroborated by the consistency of the temporal trends in moss with those of emission data for some elements [12]. The main aim of the European Moss Survey is to determine spatial differences and temporal changes in the atmospheric deposition of PTEs, estimated by their concentrations in moss. Therefore, it is of paramount importance that these concentrations are properly evaluated in terms of deviation from reference conditions, i.e., that the magnitude of pollution phenomena can be clearly depicted. As environmental quality standards (EQSs) set by legislation for the concentration of PTEs in biomonitors are missing, the interpretation of PTEs' contents in moss requires the estimation of deviation from an unaltered reference (background) condition. Therefore, the first step in any biomonitoring survey should be the definition of appropriate background values to be used as reference. In a second step, this reference can be used to calculate the extent of the deviation from this background condition.

In this paper we aimed to estimate background values for some PTEs, namely Cd, Cu, Pb, and Zn accumulated in moss collected at remote sites of Switzerland in the framework of the European Moss Survey. These background values could be used as reference for the assessment of spatial and temporal pollution phenomena. Bioaccumulation differences between moss species, although sometimes reported as important [13], being also still unclear if this variation is species- or habit-specific, were outside the scope of this paper and were not considered.

2. Materials and Methods

2.1. Selection of Background Sites and Moss Sampling

Sampling sites for the European Moss Survey were evenly distributed across the five biogeographic regions of Switzerland: Jura, Plateau, and Northern, Central, and Southern Alps, which differ in elevation, geology, and meteorological conditions as well as in flora, fauna, and population density (Figure 1). Ten remote sites were selected as pristine areas in order to estimate the background deposition. All these sites were situated in alpine regions and were under the influence of only a very modest human activity. Therefore, these sites can be regarded as representative of the "natural" situation of Switzerland.

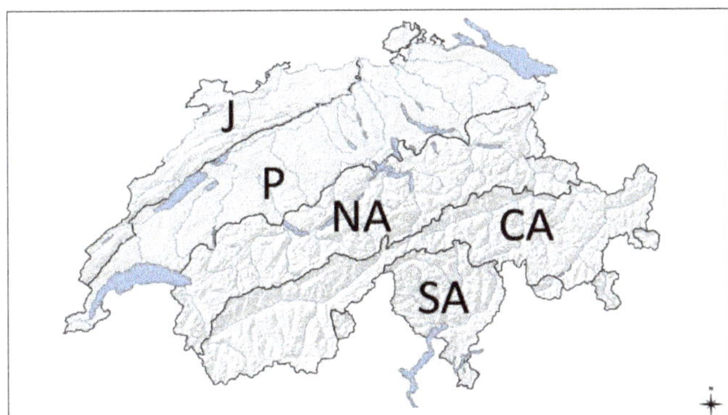

Figure 1. The five biogeographic regions of Switzerland: Jura (J), Plateau (P), Northern Alps (NA), Central Alps (CA), Southern Alps (SA). Adapted from Gutersohn [14].

The moss species *Hypnum cupressiforme* Hedw. and *Pleurozium schreberi* (Willd. Ex Brid) Mitt. were sampled, the former mostly at lowland sites and the latter mostly at Alpine sites. The moss samples were collected from tree trunks, in open areas such as forest clearings, at least 3 m away from the edge of the tree canopy, following the Moss Monitoring protocol [15]. At each site five subsamples were collected. Sampling took place every five years, namely 1990, 1995, 2000, 2005, 2010, and 2015, at the same sites and in the same period of the year (from April to October). According to the five-year cycle, the 2020 moss monitoring survey is currently ongoing, with samples being prepared for analysis and preliminary results expected by the end of 2021.

2.2. Chemical Analysis

In the laboratory, the samples were cleaned from dead or extraneous material such as litter, needles, soil, insects, etc., and only the green shoots roughly corresponding to the last three years of growth were cut for the chemical analyses. An equal amount of moss biomass was taken from each subsample and combined to form a single composite sample, taken as representative of the site following the Moss Monitoring protocol [15,16]. Samples were then dried at 40 °C. The chemical analyses were performed immediately after complete collection. Due to the time laps, the analyses were performed in different laboratories using the following analytical methods. Prior to the mineralization, each sample was pulverized in liquid nitrogen. Approximately 200 mg of moss powder was then mineralized in a microwave digestion system (Milestone Ethos 1) with 7 mL of HNO_3 and 3 mL of H_2O_2, using hermetic Teflon vessels at 130 °C and high pressure (100 bar). The mineralized and diluted solutions (up to 50 mL) were analyzed by inductively coupled plasma mass spectrometry (ICP-MS Perkin Elmer—Sciex, Elan 6100). The results are expressed on a dry weight basis (μg g^{-1} dw). Analytical quality was checked by analyzing several Standard Reference Materials: the moss standards M2 and M3 (*Pleurozium schreberi* [17,18]), BCR 61 (*Platihypnidium rapariodides*) and BCR 62 (*Olea europea*) [19], and LMS 2 (a laboratory internal moss reference material, [20]).

2.3. Statistical Analysis

For each element, the background data set was first checked for outliers using the Tukey test; in case an outlier emerged, its value was replaced by the median value of the remaining data set [21]. Based on this data set, for each element and for each year, median values and confidence limits were estimated (note that the confidence limits are not necessarily symmetric around the sample estimate, as is the case when standard errors are used to construct the confidence intervals) by bootstrapping [22]. For each element, the significance of differences between years was evaluated by means of a pairwise permutation test [23], correcting for multiple testing according to Benjamini and Hochberg [24].

3. Results and Discussion

Based on our estimates of background concentrations (Table 1), very different trends emerged for the four elements investigated. The background value of Cd showed a continuous and constant decrease, with differences requiring 10 years (two moss monitoring project sampling campaigns) to become significant. Lead showed a sharp decrease in the first two periods and from the year 2000 differences became insignificant. Copper, although with some higher values, did not show any temporal trend. Values for Zn, although decreased from 1990 to 2010, remained, overall, quite constant, without any significant difference through years.

Table 1. Median and 95% confidence limits (c.l.) of Cd, Cu, Pb, and Zn concentrations (μg g^{-1} dw) in moss at background sites of Switzerland. Different letters in a column indicate statistically significant ($p < 0.05$) differences.

	Cd		Cu		Pb		Zn	
	Median	c.l.	Median	c.l.	Median	c.l.	Median	c.l.
1990	0.20 a	0.13–0.36	4.5 ad	3.7–6.4	14.4 a	9.5–22.2	34.6	22.1–38.1
1995	0.18 a	0.12–0.28	4.8 a	4.0–6.4	5.6 b	3.2–7.3	30.9	23.0–48.6
2000	0.12 bc	0.09–0.14	7.5 bc	5.3–9.6	1.7 c	1.3–1.8	29.9	23.5–38.6
2005	0.10 cd	0.07–0.15	5.9 abcd	4.8–7.1	1.7 c	0.9–1.8	31.4	25.6–36.6
2010	0.08 de	0.06–0.10	6.4 cd	5.4–7.1	1.3 c	0.8–1.5	22.8	19.9–30.4
2015	0.07 e	0.04–0.10	4.9 ad	4.1–6.5	1.2 c	0.7–1.9	23.2	19.8–32.4
whole period	0.11	0.09–0.13	5.6	5.0–6.4	1.7	1.5–2.6	25.6	24.3–32

An important consequence of this methodological approach for estimating background element concentrations in moss is that, in addition to a proper selection of background sites, background values may change in time due to efforts to reduce emissions. Also, legal reference values for environmental pollutants measured instrumentally are updated in time, with the progress of measurement techniques, increase of knowledge, and decreasing environmental concentrations. In this light, the assessment of background values of PTEs should be intended as a dynamic process, with periodically updating with the most recent data.

To detect a significant accumulation for a given PTE, values must be significantly different from those measured at background areas. A commonly adopted approach in this sense (see, e.g., [25]) is to use the upper limit of the confidence interval as threshold. After having assessed this background threshold, a very simple and reliable method to estimate the degree of deviation from background conditions is to calculate the ratio between the concentration of a given element to its background value. This approach has the great advantage of allowing element- and species-specific differences to be overcome, thus permitting the use of a single interpretative scale in any circumstance. This method has been implemented, e.g., in the European Water Framework Directive (WFD), using so-called ecological quality ratios (EQRs), and is now successfully used is many studies [26]. Following this approach, Cecconi et al. [27] suggested an interpretative scale for Italian foliose lichens based on bioaccumulation ratios (B ratios), i.e., ratios of actual values to background values, established according to the 25th, 75th, 90th, and 95th percentiles of the frequency distributions as follows: <1 no bioaccumulation, 1–2.1 low bioaccumulation, 2.1–3.4 moderate bioaccumulation, 3.4–4.9 high bioaccumulation, >4.9 severe bioaccumulation.

On the basis of this scale, we compared the background values estimated for each period with those measured at the remaining 65 sites [12] distributed among the five Swiss biogeographic regions sampled in the corresponding year (Table 2). This comparison shows an overall absence of bioaccumulation or low bioaccumulation for all elements at all regions, with the notable exception of Pb at the Southern Alps, which showed values up to severe bioaccumulation. On an average basis, the Southern Alps were always the region with the highest bioaccumulation. However, maximum values of B ratios (data not shown) indicated a very wide array of variation, with some values reaching severe bioaccumulation for all the four investigated elements. The Southern Alps are exposed to winds from the South and, therefore, also to pollution coming from that direction (e.g., from the Po Plain, a nearby and highly polluted area of Italy), and at the same time they act as a barrier to air pollutants for the other Swiss regions.

Table 2. Bioaccumulation ratios (median values) at the five Swiss biogeographic regions during the six European moss surveys, using annual background values and the remaining 65 sampling sites. See Figure 1 for explanation of regions.

PTE	Region	1990	1995	2000	2005	2010	2015
	J	1.0	0.9	1.5	1.0	1.3	1.2
	P	0.9	0.9	1.6	1.2	1.3	1.4
Cd	NA	0.9	0.9	1.0	1.0	1.2	1.0
	CA	0.6	0.5	0.6	0.6	0.6	0.7
	SA	1.6	1.4	1.6	1.7	1.8	1.4
	J	0.7	0.7	0.4	0.7	0.6	0.7
	P	0.6	0.6	0.4	0.6	0.5	0.7
Cu	NA	0.6	0.6	0.6	0.9	0.7	0.7
	CA	0.7	0.8	0.5	0.7	0.8	0.7
	SA	1.4	1.2	0.8	1.1	0.8	1.0
	J	0.7	0.8	1.6	1.8	1.6	1.2
	P	0.7	0.8	2.0	1.9	1.3	1.3
Pb	NA	0.7	0.9	1.3	1.3	1.7	1.0
	CA	0.7	0.7	1.0	1.0	0.8	0.6
	SA	2.0	3.7	5.0	5.1	2.9	2.1
	J	0.8	0.5	0.7	0.7	0.8	0.7
	P	0.9	0.6	0.6	1.0	0.7	0.9
Zn	NA	0.8	0.7	0.8	0.9	0.8	0.7
	CA	0.9	0.6	0.6	0.8	0.8	0.8
	SA	1.8	1.2	1.2	1.5	1.3	1.2

Estimated element background values in moss for the whole period 1990–2015 (Table 1) are consistent with those estimated for the lichen *Flavoparmelia capera* (a species with ecological requirements similar to those of *H. cupressiforme* and *P. schreberi*) from Italy (Cecconi et al. 2019): $Cd = 0.18\ \mu g\ g^{-1}$ dw, $Cu = 6.2\ \mu g\ g^{-1}$ dw, $Pb = 2.4\ \mu g\ g^{-1}$ dw, and $Zn = 35.3\ \mu g\ g^{-1}$ dw. Additionally, these data also match background values estimated with a Bayesian approach in the lichen *Evernia prunastri* from Tuscany, Italy [28]: $Cd = 0.13\ \mu g\ g^{-1}$ dw, $Cu = 5.1\ \mu g\ g^{-1}$ dw, $Pb = 2.4\ \mu g\ g^{-1}$ dw, and $Zn = 22.5\ \mu g\ g^{-1}$ dw. Overall background values are also in line with the lowest values of the scales adopted to map element bioaccumulation in the European moss surveys [10]: $Cd = 0.1\ \mu g\ g^{-1}$ dw, $Cu = 4\ \mu g\ g^{-1}$ dw, $Pb = 2\ \mu g\ g^{-1}$ dw, and $Zn = 20\ \mu g\ g^{-1}$ dw.

Using these whole-period background values (upper limit of 95% confidence limits), ratios with values measured at the five Swiss biogeographic regions during the six surveys (Table 3) are indicative of temporal changes. At all regions, a gradual decrease can be seen for Cd and a marked drop for Pb; values of Zn also indicated a modest decreasing trend from 1995 to 2015, while Cu values remained quite constant. Also, in the case of temporal changes, the Southern Alps were the Swiss region with the highest bioaccumulation ratios.

Table 3. Bioaccumulation ratios (median values) at the five Swiss biogeographic regions during the six European moss surveys, using whole-period background values. See Figure 1 for explanation of regions.

PTE	Region	1990	1995	2000	2005	2010	2015
	J	1.7	1.0	0.8	0.7	0.5	0.5
	P	2.6	1.8	1.8	1.4	1.0	1.1
Cd	NA	2.6	1.9	1.1	1.1	0.9	0.8
	CA	1.7	1.0	0.7	0.7	0.5	0.5
	SA	4.4	2.9	1.8	1.9	1.4	1.1
	J	0.7	0.7	0.6	0.8	0.7	0.7
	P	0.6	0.6	0.6	0.7	0.6	0.7
Cu	NA	0.6	0.6	0.9	1.0	0.8	0.7
	CA	0.7	0.8	0.8	0.8	0.8	0.7
	SA	1.4	1.2	1.2	1.3	0.9	1.0
	J	6.2	2.3	1.1	1.2	0.9	0.8
	P	5.7	2.3	1.4	1.3	0.8	0.9
Pb	NA	5.8	2.5	0.9	0.9	1.0	0.7
	CA	5.7	2.1	0.7	0.7	0.5	0.4
	SA	17.3	10.3	3.5	3.5	1.7	1.5
	J	0.9	0.8	0.8	0.7	0.8	0.7
	P	1.1	0.9	0.8	1.2	0.6	0.9
Zn	NA	1.0	1.0	0.9	1.0	0.7	0.7
	CA	1.1	0.8	0.8	0.9	0.7	0.8
	SA	2.1	1.9	1.4	1.7	1.3	1.3

The concentrations of trace elements (C_{el}) accumulated by biomonitors can be converted into estimates of PTE deposition rates (D) [3–5,29] according to the formula:

$$D = C_{el} \cdot R \cdot t^{-1} \tag{1}$$

where R is the weight/area ratio and is the time period the sample is covering. Assuming that the final concentrations in moss represent an average equilibrium of three years with the environmental conditions at the site and knowing that pleurocarpous moss species have a weight/area ratio of 175 g m^{-2} [30], we were able to estimate annual element depositions rates at background Swiss sites as follows: Cd = 8 g km^{-2} y^{-1}, Cu = 0.4 kg km^{-2} y^{-1}, Pb = 0.2 kg km^{-2} y^{-1}, and Zn = 1.9 kg km^{-2} y^{-1}. Estimated values of Cd and Pb were consistent with the lowest annual total deposition of these elements modeled for Switzerland (Cd < 10 g km^{-2} y^{-1}, Pb < 0.28 kg km^{-2} y^{-1}) in the framework of the EMEP (European Monitoring and Evaluation Program) project, a co-operative program for monitoring and evaluation of the long-range transmission of air pollutants in Europe [31].

According to Aboal et al. [32], estimation of bulk atmospheric deposition from moss data is possible only for some PTEs such as Cd and Pb, because these elements are almost exclusively of atmospheric origin. However, for other elements this relationship is more labile since element concentrations in moss represent a steady state of non-equilibrium with the surrounding environment, rather than a time-integrated measure of element deposition [33]. Nevertheless, despite these limitations, which are intrinsic of a biological surrogate method, moss monitoring remains a very useful and efficient way of determining spatial and temporal patterns of PTEs, allowing the detection of relevant element sources and temporal trends. From this perspective, small- as well as large-scale biomonitoring surveys have, thus, their own value in regional, national, and international projects aiming at evaluating biological effects of airborne PTEs.

4. Conclusions

We statistically estimated background values for Cd, Cu, Pb, and Zn accumulated in moss from Switzerland using 10 remote sites. Our data showed that background values change over time and can be used as reference, when assessing spatial and temporal trends expressed in terms of bioaccumulation ratios with values at other sites. The use of annual background values is of great importance to identify spatial trends, while the background value over all sampling periods helps to identify temporal trends. The latter are consistent with those reported in other comprehensive, similar biomonitoring studies and require to be updated in time, possibly every five years. The use of cutoff values to be used as benchmark for bioaccumulation ratios is invaluable in having a scale for assessing ecological quality.

Author Contributions: S.L. conceived and designed the study; M.M. provided the data; Z.K. managed and reprocessed the raw data; S.L. and Z.K. analyzed the data; S.L. wrote the paper; Z.K. and M.M. supervised the text. All authors have read and agreed to the published version of the manuscript.

Funding: The moss collection and chemical analyses were funded by the Swiss Federal Office for the Environment. (FOEN).

Institutional Review Board Statement: Not applicable.

Informed Consent Statement: Not applicable.

Data Availability Statement: Raw data on moss data can be found in the publication FOEN: Bern, Switzerland, 2018.

Acknowledgments: We would like to thank the Swiss Federal Office for the Environment (FOEN) for funding the moss study in the context of the ICP Vegetation European Moss. We are grateful to the various Institutes for analyzing the moss samples: WSL, Birmensdorf, Switzerland (1990); BGR, Hannover, Germany (1995); LUFA, Hamlen, Germany (2000); VAR Tervuren, Belgium (2005, 2010); and SJI Ljublijana, Slovenia (2015). Our gratitude also goes to all the field and laboratory assistants of the various institutions for analyzing the samples and to the people who collected the mosses.

Conflicts of Interest: The authors declare no conflict of interest.

References

1. Pourret, O.; Hursthouse, A. It's time to replace the term "heavy metals" with "potentially toxic elements" when reporting environmental research. *Int. J. Environ. Res. Public Health* **2019**, *16*, 4446. [CrossRef]
2. Valko, M.; Morris, H.; Cronin, M.T.D. Metals, Toxicity and Oxidative Stress. *Curr. Med. Chem.* **2005**, *12*, 1161–1208. [CrossRef]
3. Loppi, S.; Corsini, A.; Paoli, L. Estimating environmental contamination and element deposition at an urban area of Central Italy. *Urban Sci.* **2019**, *3*, 76. [CrossRef]
4. Loppi, S. Lichens as sentinels for air pollution at remote alpine areas (Italy). *Environ. Sci. Pollut. Res.* **2014**, *21*, 2563–2571.
5. Loppi, S.; Pirintsos, S.A. Epiphytic lichens as sentinels for heavy metal pollution at forest ecosystems (central Italy). *Environ. Pollut.* **2003**, *121*, 327–332. [CrossRef]
6. Keddy, P.A. Biological monitoring and ecological prediction: From nature reserve management to national state of the environment indicators. In *Monitoring for Conservation and Ecology*; Goldsmith, F.B., Ed.; Chapman & Hall: London, UK, 1991; pp. 249–267.
7. Rühling, A.; Tyler, G. An ecological approach to the lead problem. *Bot. Not.* **1968**, *121*, 321–342.
8. Beeby, A. What do sentinels stand for? *Environ. Pollut.* **2001**, *112*, 285–298. [CrossRef]
9. Rühling, A. *Atmospheric Heavy Metal Deposition in Europe—Estimation Based on Moss Analysis*; NORD 1994:9; Nordic Council of Ministers: Copenhagen, Denmark, 1994.
10. Harmens, H.; Mills, G.; Norris, D.A.; Sharps, K. Twenty eight years of ICP Vegetation: An overview of its activities. *Ann. Bot.* **2015**, *5*, 31–43.
11. Boquete, M.T.; Aboal, J.R.; Carballeira, A.; Fernandez, J.A. Do mosses exist outside Europe? A biomonitoring reflection. *Sci. Total Environ.* **2017**, *593–594*, 567–570.
12. FOEN—Federal Office for the Environment. *Deposition of Atmospheric Pollutants in Switzerland*; FOEN: Bern, Switzerland, 2018.
13. Carballeira, A.; Couto, J.A.; Fernandez, J.A. Estimation of background levels of various elements in terrestrial mosses from GAlicia (NW Spain). *Water Air Soil Pollut.* **2002**, *133*, 235–252. [CrossRef]
14. Gutersohn, H. Naturräumliche Gliederung. In *Atlas Der Schweiz*; Eidg. Landestopographie: Wabern-Bern, Switzerland, 1973.
15. ICP Vegetation. Heavy Metals, Nitrogen and POPs in European Mosses: 2015 Survey. Monitoring Manual. 2015. Available online: http://icpvegetation.ceh.ac.uk (accessed on 27 July 2015).
16. Pennock, D.J.; Appleby, P.G. Site selection and sampling design. In *Handbook for the Assessment of Soil Erosion and Sedimentation Using Environmental Radionuclides*; Zapata, F., Ed.; Springer: Dordrecht, The Netherlands, 2002. [CrossRef]

17. Harmens, H.; Norris, D.A.; Steinnes, E.; Kubin, E.; Piispanen, J.; Alber, R.; Aleksiayenak, Y.; Blum, O.; Coşkun, M.; Dam, M.; et al. Mosses as biomonitors of atmospheric heavy metal deposition: Spatial patterns and temporal trends in Europe. *Environ. Pollut.* **2010**, *158*, 3144–3156. [CrossRef] [PubMed]

18. Steinnes, E.; Rühling, Å.; Lippo, H. Reference materials for large-scale metal deposition surveys. *Accredit. Qual. Assur.* **1997**, *2*, 243–249. [CrossRef]

19. Colinet, E.; Griepink, B.; Muntau, H. *Certification of the Contents of Cadmium, Copper, Manganese, Mercury, Lead and Zinc in Two Plant Materials of Aquatic Origin (BCR Numbers 60 and 61) and in Olive Leaves (BCR Number 62)*; European Commission: Brussels, Belgium, 1982.

20. Siewers, U.; Herpin, U. *Schwermetalleinträge in Deutschland—Moos-Monitoring 1995/96*; BGR: Hannover, Germany, 1998.

21. Tukey, J.W. *Exploratory Data Analysis*; Addison-Wesley: London, UK, 1977. [CrossRef]

22. DiCiccio, T.J.; Efron, B. Bootstrap confidence intervals. *Stat. Sci.* **1996**, *11*, 189–228.

23. Good, P. *Permutation, Parametric and Bootstrap Tests of Hypotheses*, 3rd ed.; Springer: Berlin/Heidelberg, Germany, 2005.

24. Benjamini, Y.; Hochberg, Y. Controlling the false discovery rate: A practical and powerful approach to multiple testing. *J. R. Stat. Soc.* **1995**, *57*, 289–300. [CrossRef]

25. Micó, C.; Peris, M.; Recatalá, L.; Sánchez, J. Baseline values for heavy metals in agricultural soils in an European Mediterranean region. *Sci. Total Environ.* **2007**, *378*, 13–17. [CrossRef]

26. Spatharis, S.; Tsirtsis, G. Ecological quality scales based on phytoplankton for the implementation of Water Framework Directive in the Eastern Mediterranean. *Ecol. Indic.* **2010**, *10*, 840–847. [CrossRef]

27. Cecconi, E.; Fortuna, L.; Benesperi, R.; Bianchi, E.; Brunialti, G.; Contardo, T.; Di Nuzzo, L.; Frati, L.; Monaci, F.; Munzi, S.; et al. New interpretative scales for lichen bioaccumulation data: The Italian proposal. *Atmosphere* **2019**, *10*, 136. [CrossRef]

28. Loppi, S. On the way of expressing bioaccumulation data: Does the choice of the metric really determine the outcome? In Proceedings of the 33rd Task Force Meeting of the ICP Vegetation, Riga, Latvia, 27–30 January 2019; Volume 37.

29. Contardo, T.; Vannini, A.; Sharma, K.; Giordani, P.; Loppi, S. Disentangling air pollution sources in complex urban areas by lichen biomonitoring. A case study in Milan (Italy). *Chemosphere* **2020**. [CrossRef]

30. Zechmeister, H. Annual growth of four pleurocarpous moss species and their applicability for biomonitoring heavy metals. *Environ. Monit. Assess.* **1998**, *52*, 441–451. [CrossRef]

31. Travnikov, O.; Batrakovam, N.; Gusev, A.; Ilyin, I.; Kleimenov, M.; Rozovskaya, O.; Shatalov, V.; Strijkina, I.; Aas, W.; Breivik, K.; et al. Assessment of Transboundary Pollution by Toxic Substances: Heavy Metals and POPs. Available online: https://emep.int/publ/reports/2020/EMEP_Status_Report_1_2020.pdf (accessed on 28 January 2021).

32. Aboal, J.R.; Fernandez, J.A.; Boquete, T.; Carballeira, A. Is it possible to estimate atmospheric deposition of heavy metals by analysis of terrestrial mosses? *Sci. Total Environ.* **2010**, *408*, 6291–6297. [CrossRef]

33. Gjengedal, E.; Steinnes, E. Uptake of metal ions in moss from artificial precipitation. *Environ. Monit. Assess.* **1990**, *14*, 77–87. [CrossRef] [PubMed]

Article

Analysis of Spatial Data from Moss Biomonitoring in Czech–Polish Border

Aneta Svozilíková Krakovská [1,2,*], Vladislav Svozilík [2,3], Inga Zinicovscaia [1,4], Konstantin Vergel [1] and Petr Jančík [1,5]

[1] Frank Laboratory of Neutron Physics, Joint Institute for Nuclear Research, Moscow Region, 141980 Dubna, Russia; zinikovskaia@mail.ru (I.Z.); verkn@mail.ru (K.V.); petr.jancik@vsb.cz (P.J.)

[2] Faculty of Mining and Geology, VSB-Technical University of Ostrava, 708 00 Ostrava-Poruba, Czech Republic

[3] Laboratory of Information Technologies, Joint Institute for Nuclear Research, Moscow Region, 141980 Dubna, Russia; vladislav.svozilik@vsb.cz

[4] Department of Nuclear Physics, Horia Hulubei National Institute for R&D in Physics and Nuclear Engineering, 30 Reactorului Str., MG-6 Bucharest-Magurele, Romania

[5] Faculty of Materials Science and Technology, VSB-Technical University of Ostrava, 708 00 Ostrava-Poruba, Czech Republic

* Correspondence: krakovska@jinr.ru; Tel.: +7-925-832-14-84

Received: 25 October 2020; Accepted: 10 November 2020; Published: 17 November 2020

Abstract: The purpose of the study was the analysis of spatial data gained by biomonitoring with the use of mosses. A partial goal was set to characterize the regional atmospheric deposition of pollutants in the air based on the results of the analyses and simultaneously verify the suitability of using mosses as an alternative for monitoring air quality in smaller industrial areas. In total, 93 samples of moss were collected and examined from the area of the Moravian–Silesian Region in the Czech Republic and the area of the Silesian Voivodship in Poland. The samples were analyzed using instrumental neutron activation analysis. Based on the analyses performed, 38 elements, which had been evaluated using principal component analysis, hierarchical clustering on principal components, factor analysis, correlation analysis, contamination factor, geoaccumulation index, enrichment factor, and pollution load index, were determined. The analyses resulted in a division of elements into a group with its concentrations close to the level of the values of the natural background and the second group of elements identified as emission likely originating from anthropogenic activity (Sm, W, U, Tb, and Th). The likely dominant source of emissions for the studied area was identified. Simultaneously, the results pointed to sources of local importance. The area of interest was divided into clusters according to the prevailing type of pollution and long-distance transmission of pollutants was confirmed.

Keywords: biomonitoring; bryophytes; atmospheric deposition; heavy metals; neutron activation analysis

---

1. Introduction

The environment as a complex and interconnected system consists of natural, artificial, and social components. Over the years, there has been observed an overall imbalance in the environment. In practice, monitoring especially of anthropogenic pollutants is a complex process. First, the sources and emissions are identified; then the risks are assessed. Critical emissions are controlled and economic aspects are integrated at the same time. Technical field measurements require equipment and are associated with high costs. Among the wide range approaches that make it possible to assess the state of the atmosphere, biological monitoring, i.e., biomonitoring, is increasingly coming to the fore. The complexity and at the same time the relative simplicity of its application make it an ideal tool for the given purpose. Compared to technical monitoring methods, it is inexpensive. Among the organisms used to monitor the state of the atmosphere, bryophytes are one of the most widely used. Mosses do not have a cuticle and root system and thus obtain nutrients in particles or solution directly from atmospheric deposition. They have good bioaccumulation ability, the concentration in the insole reflects the deposition from the atmosphere without being affected by soil concentrations [1].

Currently, the most problematic air pollutants in the urban environment are fine dust particles (Particulate matter). Particulate matter is the result of complex reactions of chemicals such as sulfur dioxide and nitrogen oxides, which are pollutants emitted from power plants, industries, and automobiles. They are compounds from trace elements such as Cd, Co, Cr, Cu, Fe, Mn, Pb, Sb and Zn [2] or Al, Sb, As, Cd, Fe, Mn, Ni, Pb, Cu, V, Zn according to [3]. According to the European Environment Agency (EEA) map representing the average annual concentrations of PM_{10} in 2016 (Figure 1), the areas with the highest airborne dust loads are Poland, within the Czech Republic, the Moravian–Silesian Region. Further north Italy, Bulgaria, Macedonia, Greece and Turkey.

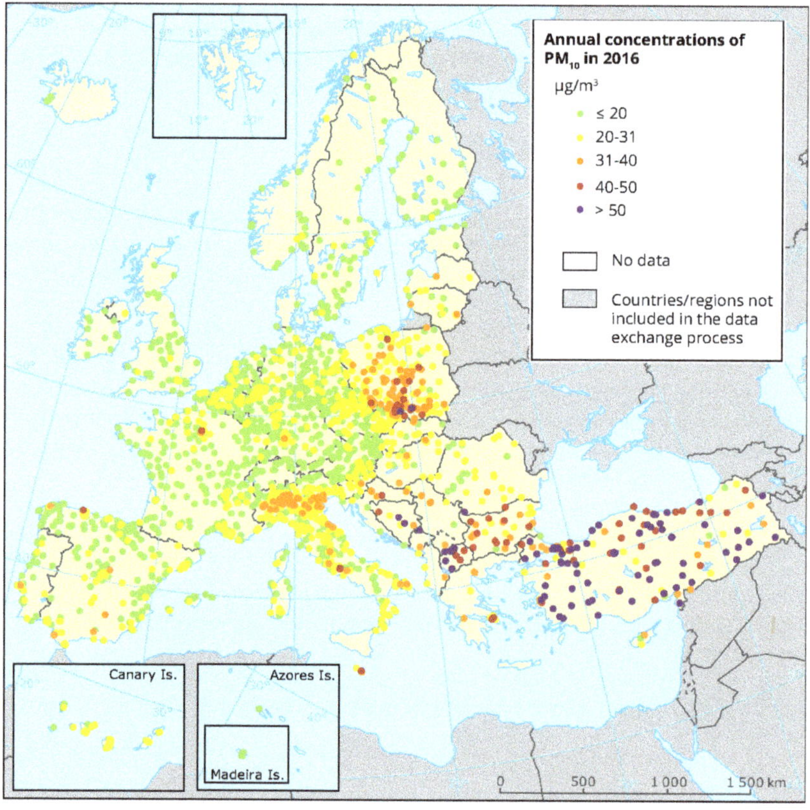

Figure 1. Annual mean concentration of PM_{10} [4].

Air quality in the Moravian–Silesian Region, especially in the Ostrava and Karviná regions, has been unsatisfactory for a long time. This fact is also evidenced by the map of average annual concentrations of PM_{10} for the Czech Republic in 2018 (Figure 2). This is mainly due to the high concentration of industry in a densely populated area, but transport and heating of households also have an impact.

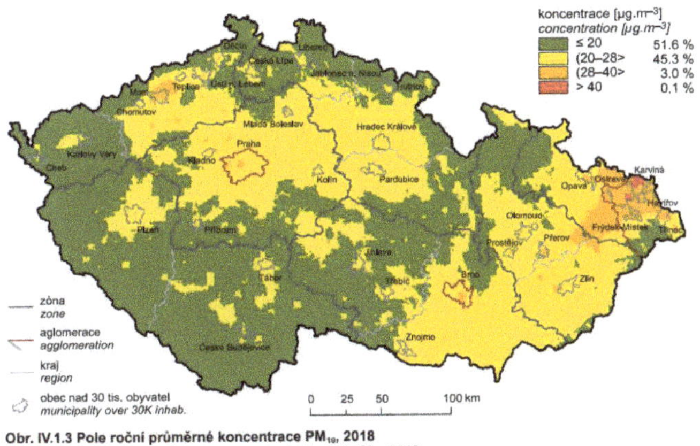

Figure 2. Annual mean concentration of PM_{10} in the Czech Republic [5].

The territory of the city of Ostrava belongs into the area of the Ostrava Basin, which is a slightly warm area with southwest and northeast wind directions. The landscape is open to the north and northeast, which causes a negative effect of northern winds in winter, but also in summer. The territory of Ostrava belongs into a moderately warm climatic region, but differs with certain peculiarities caused by a high concentration of industry, dense construction and specific conditions of the Ostrava Basin.

According to the database of the Register of Emissions and Air Polution Sources (REZZO), it is evident that primary emissions of solid pollutants are from large industrial sources, as well as local heating plants with a share of about 30%. A large share is not attributed to emissions of solids from transport. Total emissions in the Ostrava–Karviná agglomeration are approximately five times higher than in the Prague agglomeration and eight times higher than in the Brno agglomeration [6]. The Moravian–Silesian Region is characterized by the highest concentrations of $PM_{2.5}$ and PM_{10} and, compared to other regions, high values of benzo(a)pyrene appear. Directly in Ostrava, the limit value for benzene and arsenic is also repeatedly exceeded.

The aim of the paper is the analysis of spatial data obtained by biomonitoring using mosses. Furthermore, on the basis of the performed analyzes, characterization of the regional atmospheric deposition of pollutants in the air on the Czech–Polish border and verification of the suitability of the use of bryophytes as alternative monitors of air quality in small industrial areas. The another aims are identification of the significance of individual groups of sources of pollutant emissions and determination of the extent and location of polluted areas. The study evaluates the measured data and determines whether the elements contained in the samples are of natural origin, i.e., if they are part of the natural background, or whether they are of anthropogenic origin and determines whether the resulting values are of local origin or come from long-distance transmission.

The research was carried out in frame of the dissertation thesis [7].

2. Materials and Methods

2.1. Study Area

In 2015, a sampling area was selected and a network of sampling points was created on the Czech–Polish border (Figure 3), areas with the highest airborne dust loads in Europe, using GIS technologies. More species of moss were also collected from one locality for the possibility of comparing sorption capacity of different species. The assessment of the difference between the individual species is not the subject of this article. The network was also extended to locations where pollution was not expected.

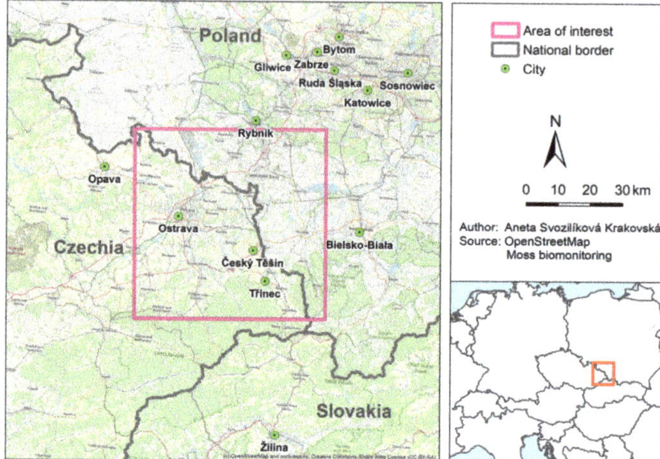

Figure 3. Map of sampling area.

2.2. Sampling and Sample Preparation

A total of 44 samples were taken from 41 sampling sites in a regular network on an area of 1600 km^2 (40 × 40 km). In 2016, the network was expanded by another 44 sampling points, where a total of 50 moss samples were taken (Figure 4). All sampling was performed according to the Monitoring Manual issued by the International Cooperative Programme (ICP) Vegetation program [8]. Sampling was performed in a square network with a side size of 60 km, with an area of 3600 km^2, where the points are a maximum of 10 km apart. A total of 85 localities were sampled, 94 moss samples were taken. According to the sampling manual, it is recommended for national monitoring to collect 1.5 samples of moss per 1000 km^2, if this is not possible, two samples in the area of 50 km × 50 km could be collected. For this research, it was sampled in a non-standard dense network, to characterize the regional atmospheric deposition of pollutants in the air. In the first campaign in 2015, the species *Hypnum cupressiforme, Brachythecium rutabulum, Hylocomium splendens, Cirriphyllum piliferum, Calliergonella cuspidata, Rhytidiadelphus squarrosus, Atrichum undulatum, Eurhinchium hians* were collected. In 2016, only *Pleurozium schreberi, Hypnum cupressiforme,* and *Brachythecium rutabulum* were collected. The necessity to use species different from recommended in the manual, is due to the dense sampling network, which extends to areas unfavorable for the growth of the recommended species, for example in city centers.

Each sampling site was selected at least 3 m from the involved treetop, preferably on the ground or surface of decaying trunks. Coarse dirt was removed in the field. According to the methodology, sampling points should be located outside urban areas, at least 300 m from main roads, villages, and industry and at least 100 m from smaller roads and houses, but these conditions could not always be met. One composite sample was prepared from each site from five to ten samples collected on an area of 50 m × 50 m. Only one species of moss was present in the composite sample. The collection was carried out in paper or plastic breathable bags. The amount of moss from one locality was one liter. Powder-free gloves were used during collection.

Samples cleaned of coarse impurities were dried in the laboratory at room temperature (20–25 °C). When handling the samples, it was necessary to avoid the use of metal aids to not contaminate the samples. The moss was cleaned of all remnants of forest bedding. Only the green parts, which represent the most frequent increments over the last three years, were taken from the sample for analysis. A sample of 6 g was weighed from the thus treated material, which was then sealed in polyethylene bags and taken to the Joint Institute for Nuclear Research (JINR) in Dubna, Russian Federation.

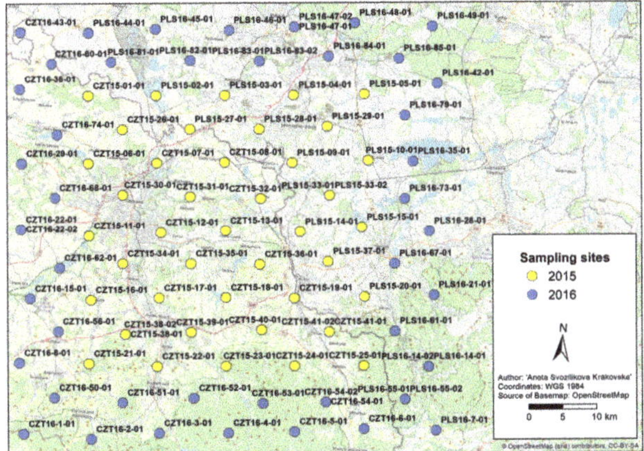

Figure 4. Map of sampling sites.

2.3. Neutron Activation Analysis

At the JINR, the sample was processed in a chemical laboratory, where 2×0.3 g of each sample was weighed. With the help of a compression piston, these two samples were compressed into sampling tablets. One was then wrapped in a polyethylene bag to determine elements with short lived isotopes (SLI) and the other in an aluminum film to determine long lived isotope (LLI). Irradiation of the sample for SLI analysis lasted 3 min and the subsequent measurement of emitted γ-radiation took place for 15 min. Irradiation of samples intended for LLI analysis lasted 3–4 days and subsequent measurement of emitted γ-radiation was performed twice, after 3 days and after 21 days for 30 and 90 min, respectively. Semiconductor high pure germanium (HPGe) detectors from CANBERRA were used for the measurement. The GENIE 2000 program was used to process γ spectra. The quality control of Neutron activation analysis (NAA) results was ensured by simultaneous analysis of the examined samples and standard reference materials (SRM) of National Institute of Standards and Technology (NIST) and Institute for Reference Materials and Measurements (IRMM): NIST SRM 1515-Apple Leaves, NIST SRM 1547-Peach Leaves, NIST SRM 1575a-Pine Needles, NIST SRM 1633b-Coal Fly Ash, NIST SRM 1633c-Coal Fly Ash, NIST SRM 1632c-Coal (Bituminous), NIST SRM 2709-San Joaquin Soil, NIST SRM 2710-Montana Soil, NIST SRM 2711-Montana Soil, SRM 2891-Copper Sand, IRMM BCR 667-Estuarine Sediment.

2.4. Statistical Data Processing

Results of NAA are in mass concentrations. Concentrations are in the form of composition data, which are defined as vectors with positive components. Components quantitatively describe parts of an entity carrying only relative information [9]. Most statistical methods are constructed assuming Euclidean geometry [10]. The special character of compositional data, different sample space, and geometry requires a different approach than standard multidimensional data, where the data are absolute. The first comprehensive approach was introduced by John Aitchison, namely log-ratio analysis [11]. Unlike other transformation methods, the identification of individual variables is impossible because it is reduced by one variable. However, to express the overall similarity between the elements measured at each site, isometric transformation is most appropriate. The basic idea is to express the composition using a suitably chosen class of transformations as real vectors and subsequent data processing is possible by common statistical methods [12].

Environmental pollution indices [13], namely Contamination factor (CF), Geoaccumulation index (Igeo), Enrichment factor (EF), and Pollution load index (PLI), make it possible to distinguish elements whose concentrations correspond to the natural background and elements whose concentrations indicate pollution on a spatial scale. The combination of these factors was used, for example, for the assessment of heavy metals levels in the sediment of the Jazmurian playa region in southeastern Iran by Shirani et al. [13] or Jiang et al. [14] in the characterization of pollution and identification of heavy metals in soil. Individual factors were used in studies of atmospheric deposition using bryophytes,, e.g., [15–19].

Statistical data processing was performed in the R studio version 3.6.1 and the statistical software STATISTICA 10. R is a freely available programming language and software environment for statistical calculations and graphics. R was created by Ross Ihaka and Robert Gentleman at the University of Auckland, New Zealand, and is developed by the R Development Core Team [20]. R contains many functions for data manipulation, calculations, and graphical outputs. Many other features are included in support packages.

To allow relative multidimensional data analysis, which requires data with Euclidean geometry, the data were transformed according to the principle of compositional data analysis. Specifically, the Isometric log-ratio transformation was used for the transformation, which allows expressing the composition in an orthonormal coordinate system. The "robComposition" package was used for transformation [21]. Exploitation analysis of data, extraction, and visualization of Principal Component Analysis (PCA) and Hierarchical Clustering on Principal Components (HCPC) were performed using the "factoextra" and "FactoMineR" packages [22]. The "corrplot" package was used to create a correlation matrix with a correlogram [23]. Factor analysis was performed in the STATISTICA program. It is a comprehensive system containing tools for data management, analysis, visualization, and development of user applications, originally developed by StatSoft, which was gain by Dell in March 2014 [24]. Before the calculation of the factor analysis itself, the data were rotated. Varimax type rotation was performed. It is an orthogonal rotation that minimizes the number of variables that have high loads with each factor in common. It is calculated by the sum of the variances of the factor load squares in the individual columns.

2.4.1. Principal Component Analysis

The main goal is to simplify the description of a group of mutually linear dependent, i.e., correlated features. The motivation is to replace a large number of input variables with a much smaller number of new variables, so-called components, without much loss of essential information about the input data. It is a method of a linear transformation of the original characters into new, uncorrelated variables, called main components. Each main component represents a linear combination of the original features, the basic characteristic of each of them is its degree of variability, i.e., variance. The principal component method is one of the basic multi-elemental analysis used in the evaluation of atmospheric deposition using biomonitoring [25,26].

2.4.2. Factor Analysis

Factor analysis (FA) is a widely used method in evaluating the resulting concentrations from biomonitoring using bryophytes, e.g., [27–30]. Factor analysis is a multidimensional statistical method, explaining the variance of observed variables using a smaller number of potential variables—the factors. Its essence is the analysis of the structure of mutual variables based on the assumption that these dependencies are the result of a certain smaller number of background immeasurable factors. These factors are called common factors. The goal is to reduce the number of variables and reveal the structure of the relationship between variables. Factor analysis can to some extent be considered an extension of the principal component method (PCA), but unlike PCA, it is based on an attempt to explain the relationship between variables. The shortcomings of PCA include that it is dependent on changes in the scale of variables. The factor analysis approach makes it possible to eliminate this shortcoming. The weaknesses of factor analysis lie in the ambiguity of the estimation of factor parameters (i.e., the dependence of the FA result on the rotation used) and in the need to specify the number of common factors before performing the analysis. The advantages of FA are greater economy and generality. Like PCA, the problem is the interpretation of the factor if the variables do not have a multidimensional normal distribution. The main goal of PCA is to explain the maximum variability of data, the main goal of FA is to explain the covariance between variables. The basis of factor analysis is the assumption that the observed covariances between variables are the result of the action of common factors and not the relationship between variables. Thus, the FA assumes that variables are a linear combination of hypothetical variables-factors [31–33].

2.4.3. Contamination Factor

Contamination factor was first used by Hakanson [34] to study the pollution of reservoirs and to determine which ones need more attention in terms of potential environmental risk. To evaluate the degree of contamination of bryophytes used for air biomonitoring, the use of CF for the first time described by Fernandez et al. [35]. According to these authors, a scale was used to categorize individual sampling points. The contamination factor evaluates the degree of contamination for individual elements, but also the degree of contamination for individual sampling points. This allows you to compare data from different regions. It is calculated as the ratio of the concentration of the element in the moss and the background value of the element in the studied area.

$$CF = \frac{\text{concentration}}{\text{background}} \tag{1}$$

2.4.4. Geoaccumulation Index

The Geoaccumulation Index (I_{geo}) was originally introduced by Müller [36] to determine and define metal contamination in sediments by comparing current concentrations with pre-industrial values. It is calculated according to the formula [37]:

$$I_{geo} = log_2 \left[\frac{c_n}{1.5 B_n} \right] \tag{2}$$

c_n concentration in moss sample *n*,
B_n background value for moss sample *n*,
a factor of 1.5 is used due to possible variability in background values.

2.4.5. Enrichment Factor

Given that some authors, e.g., Covelli and Fontolan [38] criticize the use of Igeo to assess the degree of contamination and suggest the use of normalized values, the present article also used a method of comparing the ratio of individual metals to a reference element that is probably of geogenic origin, and the impact of metals on the environment. The degree of pollution was determined using the Enrichment factor (EF). EF is a standardized method proposed by Sinex and Helz [39] to assess metal concentrations. The concentration of metals is usually normalized as a ratio to another component of the sediment. Rubio et al. [40] stated that there is no general agreement on the most suitable sediment component to be used for normalization. The most usable are Al, Fe, total organic carbon, and sediment grain size. However, an important criterion is the non-anthropogenic origin of the element. The component chosen for this purpose should not be variable depending on anthropogenic activity [41]. For this reason, the use of the given elements as comparative elements was excluded, as their occurrence in the examined area may be influenced by the presence of the metallurgical industry. Therefore, the scandium element was selected for comparison, which has a relatively small application in industry, and although it reached high values in the calculation of the contamination and geoaccumulation factor, this was probably due to swirling dust and contamination of samples with soil particles. The crustal origin of scandium is confirmed by the results of a biomonitoring study such as Olise et al. [42]. Several studies have been performed using scandium as a reference element [43,44]. The formula for calculating EF is:

$$EF = \frac{\left(\frac{metal}{Sc}\right) sample}{\left(\frac{metal}{Sc}\right) background} \tag{3}$$

where $\left(\frac{metal}{Sc}\right)_{sample}$ is the ratio of metal to concentration of Sc in the sample and $\left(\frac{metal}{Sc}\right)_{background}$ is the ratio of metal to concentration of Sc in the reference samples.

2.4.6. Pollution Load Index

The Pollution Load Index (PLI) was designed and tested to assess the heavy metal pollution of estuaries. Tomlinson [45] stated that although there are many indicators to facilitate the detection of heavy metal pollution, there are significant problems in evaluating the heavy metal load on rivers. The interpretation of the results is complicated by differences in the composition of species and conditions in different places, differences in the period of sampling, or the age of organisms and methods of storing metals in the bodies of organisms. For the biomonitoring of air pollutants using mosses and lichens, PLI was used, for example, by Salo [46], who studied the relationship between the concentration of anthropogenic magnetic particles and heavy metals and compared the correlations between magnetic sensitivity and PLI. The Tomlinson contamination index [47] indicates the extent to which a sample exceeds the concentrations of heavy metals in the natural environment and indicates the overall state of toxicity for the sample [46]. PLI is defined as the square root of the product of the contamination factors for a given sample:

$$PLI = \sqrt[n]{CF1 \times CF2 \times CF3 \times ...CFn} \tag{4}$$

where *CF* indicates contamination factors for a single site sample.

3. Results

In 2015, a total of 44 samples were taken from 41 sites. In 2016, another 50 samples were taken from 44 localities. Concentrations of 38 elements were obtained for samples from 2015, and 47 elements were evaluated in 2016. Due to incomplete data, given the uncertainties in the neutron activation analysis, the elements, Yb, Lu, Hg, Cu, Zr, In, Eu, Gd, and Dy, were removed from further processing. The sampling point CZT-15-13-01 was also removed, as it was collected and analyzed separately and the resulting concentrations and composition of the elements are different from other sampling points. 93 samples were entered into statistical analyzes and 38 variables (elements) were evaluated. The basic numerical characteristics of the analyzed data are given in the (Table 1).

The values of the coefficient of variation express the inhomogeneous character of the data set. The skewness coefficient for some elements (e.g., Cr, Fe, Zn, As, Cd, W, and Au) indicates a positive skewness of the data set, this fact is also evidenced by higher values of averages than the median values. Based on the results of the sharpness coefficient, it can be concluded that the distribution is more pointed than normal. These facts are also confirmed by frequency histograms. It is evident from the histograms that the data for most elements do not have a normal or Gaussian distribution according to the author Carl Friedrich Gauss.

To allow relative multivariate data analysis, which requires Euclidean geometry data, a nonlinear data transformation was performed according to the principle of compositional data analysis. Specifically, the Isometric log-ratio transformation was used for the transformation.

Table 1. Descriptive statistics of measurement, $n = 93$.

	Min [mg/kg]	Max [mg/kg]	Mean [mg/kg]	Median [mg/kg]	std.dev [mg/kg]	var [mg/kg]^2	var. koef. /	Skew /	kurt. /
Na	84.9	1150	276	224	173	29,869	0.6	2	9.1
Mg	789	4790	2525	2340	1009	1,017,205	0.4	0.3	2.1
Al	305	11,000	2530	1830	2119	4,488,235	0.8	1.4	4.9
Cl	69.3	2240	502	389	394	154,875	0.8	1.9	7.7
K	4660	20,200	11,024	11,100	3104	9,634,437	0.3	0.1	3
Ca	1540	10,600	5941	5940	2411	5,811,460	0.4	0	2
Sc	0.053	1.86	0.51	0.41	0.384	0.147	0.8	1.1	3.8
Ti	22.3	923	202	133	177	31,182	0.9	1.5	5.2
V	0.551	14.9	4	2.88	3.13	9.78	0.8	1.2	4.1
Cr	1.1	34.1	6.84	5.41	5.56	30.96	0.8	2.1	8.9
Mn	45.7	767	230	195	153	23,449	0.7	1.5	5
Fe	338	18,700	2203	1680	2258	5,099,947	1	4.5	31.9
Co	0.119	2.13	0.76	0.63	0.483	0.233	0.6	1.1	3.6
Ni	0.711	8.26	2.98	2.69	1.465	2.15	0.5	1.2	4.8
Zn	30.6	587	111	85.4	106	11,295	1	3.3	13.6
As	0.286	3.75	1.07	0.97	0.518	0.269	0.5	2.1	10
Se	0.06	2.39	0.64	0.39	0.537	0.289	0.8	1.4	3.8
Br	1.65	7.73	3.54	3.27	1.31	1.72	0.4	1	3.9
Rb	5.94	63.8	17.2	13.3	11.5	132	0.7	1.9	6.8
Sr	5.65	69.2	27.4	26.6	13.3	177	0.5	0.5	3
Mo	0.016	1	0.39	0.34	0.245	0.06	0.6	0.6	2.3
Cd	0.02	7.09	1.18	0.57	1.719	2.95	1.5	2.2	6.8
Sb	0.049	1.31	0.34	0.28	0.226	0.051	0.7	2	7.8
I	0.349	4.08	1.41	1.31	0.724	0.525	0.5	1	4.2
Cs	0.078	1.74	0.41	0.33	0.282	0.079	0.7	1.7	7.4
Ba	13.1	209	63.1	55.8	33.8	1145	0.5	1.3	5.8
La	0.194	6.13	1.6	1.24	1.211	1.47	0.8	1.2	4.1
Ce	0.011	15.6	3.23	2.43	2.682	7.19	0.8	1.6	6.6
Nd	0.249	6.82	1.91	1.74	1.333	1.78	0.7	1	4.1
Sm	0.026	1.03	0.26	0.2	0.201	0.04	0.8	1.3	4.5
Tb	0.003	0.16	0.04	0.03	0.03	0.001	0.8	1.3	4.5
Tm	0.002	0.08	0.02	0.02	0.017	0	0.7	1.2	4.2
Hf	0.025	1.56	0.39	0.27	0.34	0.115	0.9	1.2	3.9
Ta	0.004	0.21	0.05	0.04	0.041	0.002	0.8	1.3	4.4
W	0.03	1.38	0.27	0.23	0.233	0.054	0.9	2.2	9.9
Au	0	0.07	0	0	0.008	0	4.4	7.4	59.4
Th	0.049	1.92	0.49	0.38	0.4	0.16	0.8	1.3	4.5
U	0.021	0.56	0.18	0.14	0.14	0.02	0.8	1	3.3

3.1. Principal Component Analysis

Scree graph analysis, i.e., a bar graph of eigenvalues, was used to identify the number of major components, where the break point from rapid descent to gradual determines the number of "useful" components. The useful components are thus separated by a distinct break point, and the x coordinate of this break is the index value sought. In this case, two. The biplot is a natural consequence of the singular value decomposition of a matrix [48]. The Scatterplot (Figure 5) shows the component weights for the two main components and allows to compare the distances between the variables. Short distance means a strong correlation. Sampling points located far from the origin of coordinates are extremes (CZT15-11-01, CZT15-03-01, PLS16-45-01, PLS16-83-01), the closest, on the other hand, are the most typical. The sites located in the diagram close to each other are similar, far from each other are different. Sites located clearly in one cluster are similar, yet dissimilar to objects in other clusters. Isolated sites (PLS16-83-01) may be strongly dissimilar to others unless there is an apparent inhomogeneity due to skewed data. The color scale distinguishes the quality of the data representation in the graph.

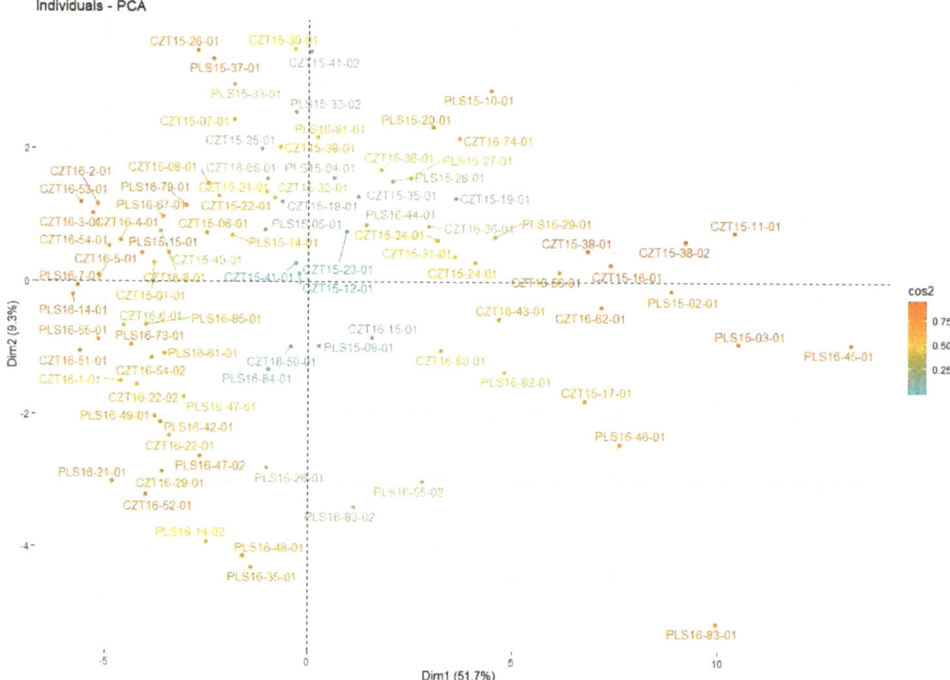

Figure 5. Scatterplot of the component score of individual sampling sites.

The Plot Components Weight (Figure 6) shows the component weights for the first two main components. The distances between the variables are compared, shorter means a strong correlation. Variables close to the origin are of little importance (Au). Variables with an angle of 0° between their guides are completely positively correlated. The group of elements U, Sc, Ta, Al, Co, Sm, Ce, La, V, Th, Tm, Tb have a strong correlation with each other. Also the elements Fe, Cr, Na, Sr, Nd, Ni, Sb and Cs are correlated with the mentioned group, but they are no longer of such significance. Variables with an angle of 90° are completely uncorrelated and variables with an angle of 180° are negatively correlated, for example Au, Cl, K to Rb, Cd, Se, Mn. The color scale indicates the level of contribution of the element to the main component.

A principal component analysis was performed for the three major components. Subsequently, hierarchical clustering was performed on these three components. The individual sites are visualized on the graph of the main components (Figure 7) and the individual clusters are color-coded. The map shows a better division of the sampling points into clusters (Figure 8).

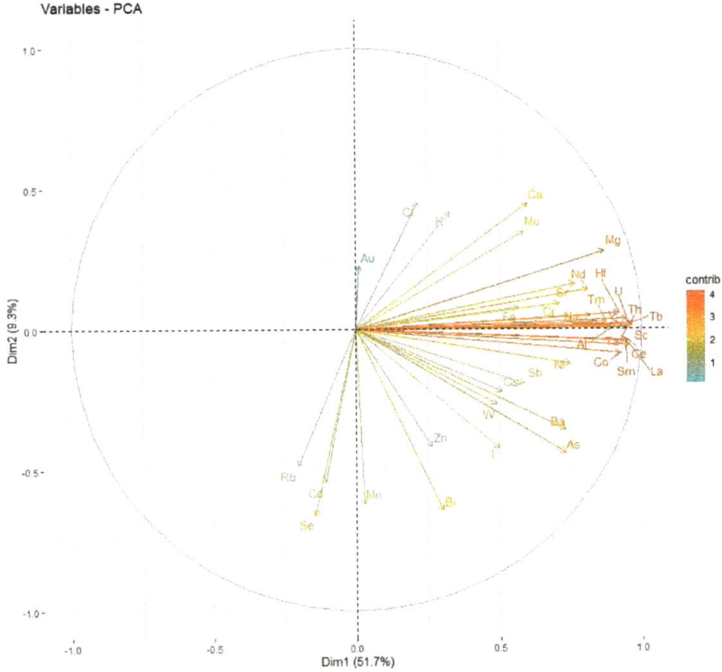

Figure 6. Plot Components Weigh of individual elements.

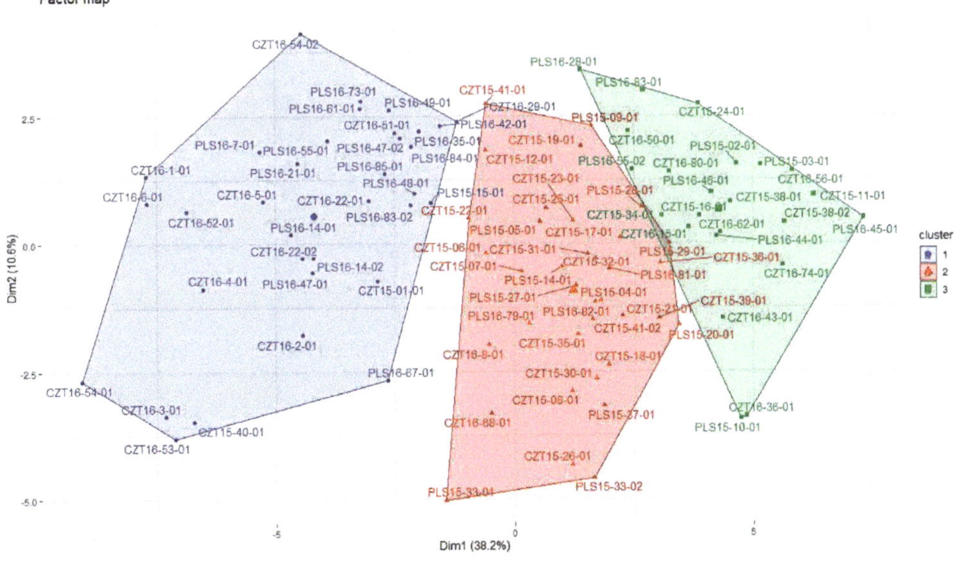

Figure 7. Graph of main components with clusters.

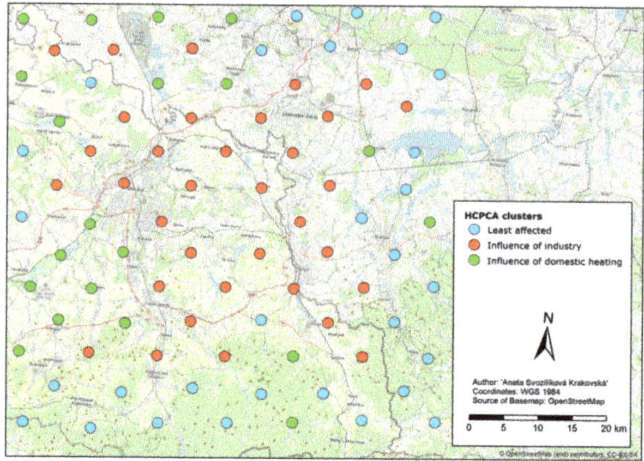

Figure 8. The result of hierarchical clustering on three components.

3.2. Factor Analysis

Factor analysis was performed in the software STATISTICA. The optimal number of factors to be retained was determined by plotting line rubble-Scree plot [49] eigenvalues of all factors. The ideal number of factors can be determined at the point in the graph where the highest decrease in eigenvalues is between the two factors. The biggest break is between one and two factors and another only between the fifth and sixth factors, so the factor analysis was performed for five factors.

Prior to the analysis itself, the data were rotated to clearly distinguish the load pattern, i.e., to differentiate the factors indicated by the high load for some variables and low load for others. Namely, Varimax normalized rotation was used. As a result, those variables (elements) whose load was greater than 0.6 for a given factor were selected (Table 2, highlighted in red).

Table 2. Factor loads of individual elements, $n = 93$.

	Factor 1	Factor 2	Factor 3	Factor 4	Factor 5
Na	0.79	0.16	0.16	0.07	0.13
Mg	0.76	0.07	0.37	0.05	0.39
Al	0.94	0.04	0.13	−0.06	0.07
Cl	0.11	−0.02	0.12	0.08	0.82
K	0.25	0.07	0.00	0.22	0.79
Ca	0.37	0.01	0.72	0.03	0.42
Sc	0.96	0.11	0.19	0.01	0.03
Ti	0.92	0.03	0.08	−0.09	0.11
V	0.91	0.08	0.25	−0.07	0.03
Cr	0.56	0.13	0.62	0.03	−0.13
Mn	−0.08	0.61	−0.12	−0.33	−0.03
Fe	0.38	0.19	0.65	0.09	−0.17
Co	0.85	0.26	0.30	−0.01	−0.04
Ni	0.67	0.39	0.18	0.19	−0.08
Zn	0.07	0.81	0.00	0.13	0.10
As	0.65	0.62	0.00	0.03	−0.13
Se	−0.08	0.09	−0.17	−0.74	−0.37
Br	0.20	0.63	−0.14	−0.45	0.11
Rb	−0.12	0.41	−0.55	0.17	−0.26
Sr	0.58	0.22	0.48	−0.03	0.35
Mo	0.36	0.17	0.70	0.29	0.13

Table 2. *Cont.*

	Factor 1	Factor 2	Factor 3	Factor 4	Factor 5
Cd	−0.09	0.09	−0.16	−0.82	−0.01
Sb	0.36	0.55	0.48	0.10	−0.13
I	0.28	0.58	0.39	−0.33	−0.05
Cs	0.47	0.36	0.09	0.24	−0.33
Ba	0.61	0.59	0.05	−0.13	0.17
La	0.95	0.18	0.12	−0.06	0.10
Ce	0.96	0.13	0.11	−0.04	0.06
Nd	0.83	−0.08	0.19	−0.05	0.12
Sm	0.96	0.09	0.10	−0.10	0.07
Tb	0.97	0.10	0.15	0.00	0.07
Tm	0.89	0.14	0.05	0.12	0.08
Hf	0.91	0.12	0.13	0.03	0.15
Ta	0.96	0.09	0.14	−0.01	0.05
W	0.45	0.02	0.27	−0.62	−0.08
Au	−0.05	−0.14	0.30	0.04	0.02
Th	0.96	0.11	0.17	0.03	0.05
U	0.87	0.21	0.32	0.05	0.07
Expl.Var	17.06	3.71	3.70	2.42	2.34
Prp.Totl	0.45	0.10	0.10	0.06	0.06

3.3. Correlations Analysis

A correlation matrix was created to assess the strength of the linear relationship between the quantities (Figure 9). Positive correlations are shown in blue and negative correlations in red. The color intensity and brand size are directly proportional to correlation coefficients. The color legend on the right shows the correlation coefficients and their corresponding color.

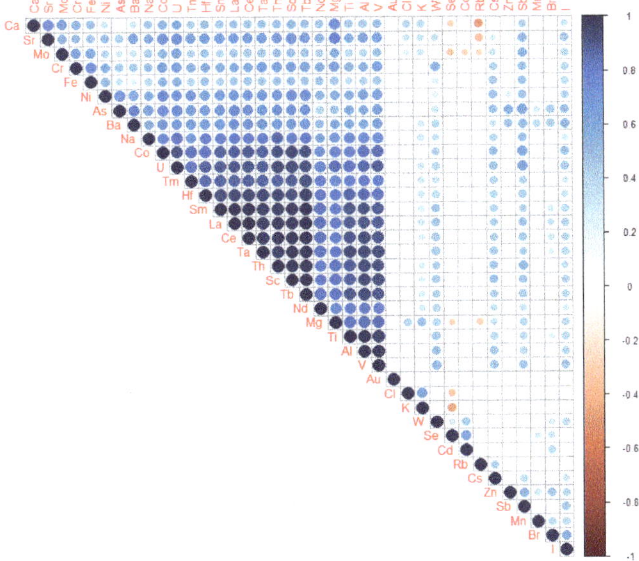

Figure 9. Correlation matrix.

The correlation matrix shows a high correlation between the elements Na, Co, U, Tm, Hf, Sm, La, Ce, Ta, Th, Sc, Tb, Nd, Mg, Ti, Al, and V. A weaker positive correlation is observed for Ca, Sr, Mo, Cr, Fe, Ni, As and Ba. Negative correlation with most elements is recorded for Se, Cd, and Rb, they are positively correlated with each other. Other elements do not show significant correlations. Based on the results of the correlation matrix, clustering took place. A cluster is a group of objects whose distance is less than the distance with objects that do not belong to the cluster. Clustering was performed according to Wald's criterion, where the principle is to minimize cluster heterogeneity according to the criterion of minimum increment of intragroup sums of deviations of objects from the center of gravity of clusters [50]. The result is a dendrogram (Figure 10). When dividing the elements into three clusters, Rb, Se, Cd, Mn, Zn, Br, Au, Cl, and K appear in the first group. The second cluster contains V, Al, Ti, Co, U, Hf, Sm, La, Ce, Ta, Th, Sc, and Tb, and the third cluster Nd, Na, Tm, Mg, Sr, Ca, Mo, Cr, Fe, Ni, As, Ba, W, Cs, Sb and I.

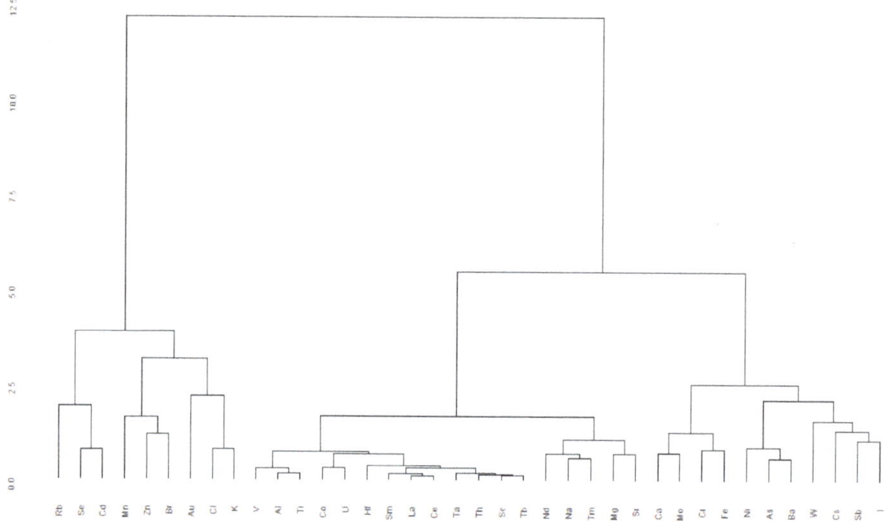

Figure 10. Hierarchical clustering based on a correlation matrix.

3.4. Contamination Factor

The Fernández and Carballeira [35] scale was used to assess contamination by the contamination factor (CF), which allows the categorization of sampling points for each element (Table 3). The proposed scale includes six categories, from a contamination factor value less than 1 (no contamination) to values greater than 27 (extreme contamination). Data from moss collections in 2015 and 2016, but also from the biomonitoring campaign in 2017 (a total of 337 samples) were used to calculate background values. The concentrations of the individual elements were sorted from the smallest to the largest, and each sampling point was assigned an order for each element. The sum of the order of the individual elements was summed for each sampling point and the resulting sums were sorted again. The arithmetic mean of the first 5% points (17 values) was used as the background value.

$$CF = \frac{Med(concentration)}{background} \tag{5}$$

Table 3. Classification of elements according to the degree of contamination.

No Contamination	Suspected	Slight	Moderate
Mn	As	Al	Sm
	Cd	Cr	W
	Ni	Fe	U
	Sb	V	Tb
	Ba	Zn	Th
	Sr	Co	Sc
	Se	Mo	
	Ca	Ce	
	K	Na	
	Mg	Cs	
	Rb	Nd	
	Ba	Ti	
	Au	Cl	

3.5. Geoaccumulation Index

The resulting values of the geoaccumulation index (Igeo) were classified by Muller [36] as shown in the following Table 4.

Table 4. Degree of metal pollution in terms of seven classes.

I_{geo} Class	I_{geo} Value	Classification
0	<0	uncontaminated
1	0–1	uncontaminated to moderately contaminated
2	1–2	moderately contaminated
3	2–3	moderately to strongly contaminated
4	3–4	strongly contaminated
5	4–5	strongly to extremely contaminated
6	>5	extremely contaminated

The resulting Igeo element distribution is shown in the table (Table 5).

Table 5. Classification of elements according to the degree of contamination.

Uncontaminated	Uncontaminated to Moderately Contaminated	Moderately Contaminated
Ni	As	Al
Mn	Cd	Fe
Se	Cr	Ce
Ca	Sb	Sm
Rb	V	W
Au	Zn	U
Br	Ba	Nd
I	Sr	Tb
	Co	Th
	Mo	Sc
	K	
	Mg	
	Na	
	Cs	
	Ba	
	Cl	
	Ti	

3.6. Enrichment Factor

Due to the different scales reported in different publications [51,52] and due to the use of Sc as a normalizing element, the elements were divided into categories according to value. According to Kłos et al. [44], elements with an enrichment factor (EF) value < 1 indicate elements originating from the natural background and elements with an EF value > 1 elements affected by anthropogenic activity (Table 6).

Table 6. Classification of elements according to the degree of contamination.

EF < 1		EF > 1
Mn	Sr	La
Rb	Zn	Hf
Se	Na	Fe
Au	Cs	Th
Ca	Co	Ta
Ni	Mo	Tb
Br	Tm	Sm
K	Ti	U
I	Cr	W
As	Cl	
Mg	V	
Ba	Nd	
Cd	Ce	
Sb	Al	

3.7. Pollution Load Index

Tomlinson et al. [45] proposed a scale for evaluating the results, where zero means a perfectly clean site, a value of one means the presence of pollutants at the baseline level and concentrations above one, show a gradual deterioration in the quality of the investigated environment. Thus, it can be stated that a value of PLI < 1 defines elemental pollution close to the background value, while values of PLI > 1 indicate pollution of a given locality. Since almost the entire study area shows PLI values greater than one, a map was created to better present the results, dividing the resulting values into 6 categories (Figure 11).

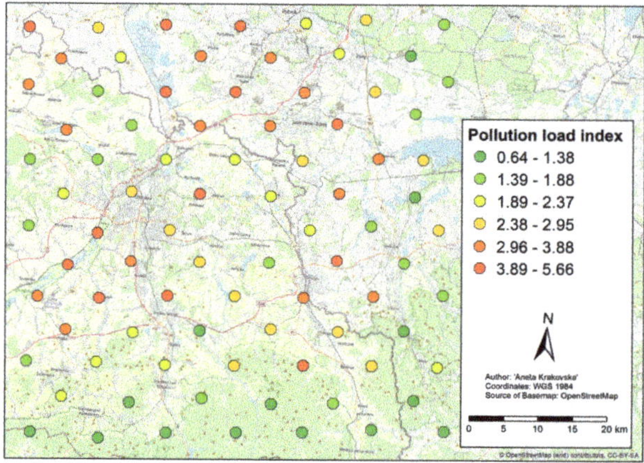

Figure 11. Pollution load index.

4. Discussion

The findings about potential sources in terms of elemental composition, presence of potential emission sources, but also climactic and geomorphologic conditions, which allow the transmission of pollution from its source to the place of deposition, were included in the evaluation. The result of the principal components analysis is represented by the dispersion diagram of the component score (Figure 5). Based on the rules explaining the dispersion diagram, a map (Figure 12) of mutually correlated areas was created.

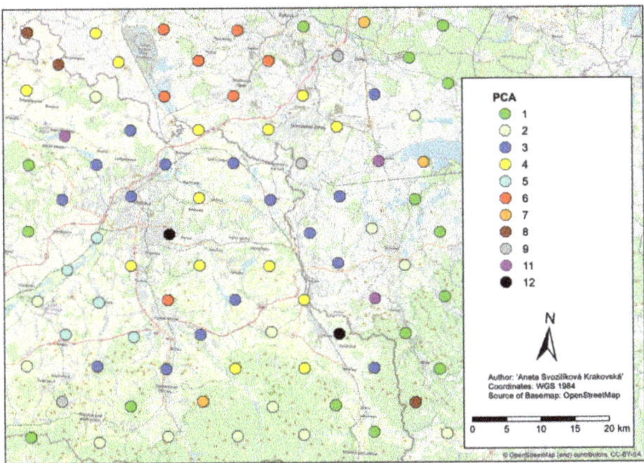

Figure 12. The result of the principal components analysis.

The points included in groups 1 and 2 are found in locations least affected by anthropogenic activity from the studied location. The sampling sites in category 3 are influenced by industrial activity and also local domestic heating and emissions from transportation, whereas the points in group 5 are not affected by emissions from industrial areas to such an extent. According to the analysis, the sites in group 6 are the most different from other sites. These points are located on the Polish side in an area of mining, engineering, and metallurgical industry. At the same time, one point is found near the town Frýdek-Místek where the metallurgical and machine-building industry is well-developed. According to the dispersion diagram, the points in category 12 are the closest to the origin and as such the most typical for the studied area; in this category, there are two points located near the village Kunčičky in the Ostrava region and close to Třinec. Metallurgical companies can be found in both of these localities.

Based on the hierarchical clustering on principal components analysis (HCPC), the sample points were divided into three clusters (Figure 8). From the clusters, we can identify areas affected by the industry, areas not significantly affected by the industry but containing a large concentration of domestic heating, and areas least affected by anthropogenic activity.

The division of elements into five factors according to the factor analysis and also the spatial distribution of the factor loadings of individual factors can be explained by the following description. Factor 1 is composed of a large number of elements (Na, Mg, Al, Sc, Ti, V, Co, Ni, As, Ba, La, Ce, Nd, Sm, Tb, Tm, Hf, Ta, Th, and U). Elements of natural origin can be influenced by the deposition from industrial sources based on the localization of the highest values of the load factor. Another source could be dust/soil particles. As described by Shetekauri et al. [28], V and Ni are the dominant elements in areas with metallurgical and mining industries. Simultaneously, the aforementioned authors attribute the natural crustal origin of elements to factor 1, which in their case also contains Ti, V, Ni, Co, As, Th, and U. Zinicovscaia et al. [27] states a combination of geogenic and anthropogenic associations in relation to Co and U as elements belonging to factor 1. Apparently, the measured results may have been affected by soil particles.

The most prominent elements in factor 2 are represented by Mn, Zn, As, and Br. Identical elements were also identified in one of the factors in the study of Olise et al. [42]. The authors note that coal impurities emitted at high temperatures contain As, Rb, As, Br, Mn, and Cr. This fact is also confirmed by other authors [53–55].

According to the graph of component weights (Figure 6), the element I can also be added to this group. All above-mentioned elements except for As do not show mutual correlations with other elements (Figure 9). In the calculation of all factors, manganese came out as the element least affected by anthropogenic activity. Both iodine and bromine have negative values of the geoaccumulation factor; as such, they can be considered elements of natural origin. Due to its origin, arsenic stands apart from the group as it enters the air practically only due to human activity, particularly by burning fossil fuels and wood conserved with arsenic-containing products. According to the Korzekwa et al. [29] who utilize the method of biomonitoring using mosses in Poland, another source of As besides burning fossil fuels can be represented by pesticides or products used for wood curing; this is because in their case, higher concentrations of As were found in forests or close to agricultural areas. Metallurgical companies processing copper, lead, and other metals containing arsenic in their ore are considered to be a source, too. However, the values of the contamination factor of arsenic reach the "moderate" category at maximum, and they only do so in two points (Figure 13).

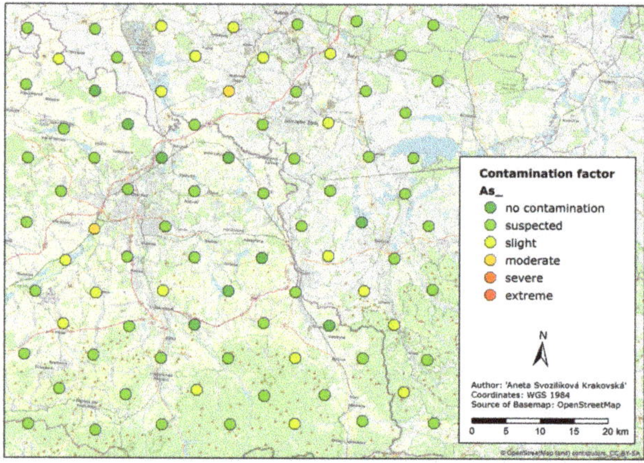

Figure 13. The distribution of values of the contamination factor for As.

One point is located in a part of Ostrava called Polanka nad Odrou where the moss was collected in a place with the following activity: Buyout and processing of alloy steel waste, processing of construction materials, selling of solid fuels and metallurgical materials, and manufacturing of asphalt mixes. The second point with its contamination factor in the "moderate" category is located in the town of Wodzisław Śląski where there are long-term issues with emissions originating from fuel burning in home boilers, undesirable technical conditions, and energetic efficiency. Other issues are also represented by the low quality of fuel burned and the burning of waste. Higher values of the contamination factor, up to the category "severe", are displayed by four points in the case of zinc. The points described can be found in the area around the municipalities Strahovice, Pszów, Marklowice, and Żory. Significant anthropogenic sources of zinc are represented by mining, zinc, lead, and cadmium refining, steelmaking, burning of coal and other organic fuels, ore mining and processing, and utilization of zinc-containing fertilizers. Zinc is utilized to a large extent (up to 40% of production) as an anti-corrosive protective material for iron and its alloys.

According to the graph of component weights (Figure 6), Ca and Mo are more correlated with each other when compared to the pair of Cr and Fe. Based on the values of the contamination factor, calcium appears as an element included in the natural background; similarly, molybdenum belongs to the essential microelements necessary for plant development. However, Korzekwa et al. [29] states the origin of molybdenum to be the metallurgical plant. In line with this statement, higher values of the contamination factor of molybdenum correspond with the distribution of residences, pointing to anthropogenic sources such as fossil fuels burning, metallurgy, but also mining and the electrotechnical industry. In the case of Ca, the drawing of the element can occur from soil substrate according to a number of sources [42,56–59].

Higher concentrations of chromium can be found in the regions of Ostrava, Třinec, and the surrounding areas of the town Wodzisław Śląski where the mining, metallurgical, and engineering industries are well-developed. A contamination factor of the "moderate" category is also spread around smaller residencies and this reality can be caused by the burning of fossil fuels (in the form of Cr^{3+}). Other sources can be represented by cement plants, communal waste incinerators, exhaust gases from automobiles with a catalyst, emissions from air-conditioning cooling towers using compounds of chromium as corrosion inhibitors, and flying asbestos from automobile brakes. High values of the contamination factor of Cr and Fe are also apparent on the north side of the Beskids Protected Landscape Area near the municipalities Řeka, Vyšní Lhoty, and Komorní Lhotka where the industry concentration is not so dense but where the massive of the Moravian–Silesian Beskids begins. Therefore, pollution is likely brought by the blowing wind from the north and its fallout and capture take place on the windy side of the massive. Iron and chromium interfere and their main sources are found in the agglomeration of Ostrava/Karviná/Frýdek Místek and in the vicinity of iron foundries. A particular affinity to absorption and accumulation of dust particles was also confirmed by Pandey et al. [60] and Olise et al. [42], who also confirms Fe as the dominant element in moss samples in the vicinity of iron and steel plants.

Factor 4 is made up of elements Se, Cd, and W. On one hand, according to the geoaccumulation factor, selenium can be considered an element of geogenic origin. This is also indicated by the scatter of its higher concentrations in the regions of the Beskids Protected Landscape Area on both the Czech and the Polish sides, the Poodří Protected Landscape Area, Oderské vrchy Natural Park, and the Cysterskie Kompozycje Krajobrazowe Rud Wielkich Protected Landscape Area. On the other hand, according to the contamination factor, selenium is "suspected of contamination", which corresponds with the higher concentrations in the vicinity of the towns Rydyłtowy and Rybnik where the main emission sources of selenium are thermal power plants and plants of metallurgical and chemical industry. In the case of cadmium, the situation is questionable. The scatter of higher concentrations and higher levels of contamination factor of cadmium is particularly apparent in the peripheral parts of the studied area where clean locations without industry or dense settlement are found. Based on The Integrated pollution register [61], the anthropogenic emissions of cadmium are approximately eight times higher than the emissions from natural sources. One of the explanations of such scatter of concentrations is the long-distance transmission of emissions; in areas of immediate proximity of emission sources, the deposition of other substances prevails and cadmium is transported into more distant locations. Higher concentration could be caused also by transport. The points located in agricultural areas can be affected by using phosphate fertilizers with cadmium ingredients and loading waste treatment plant sewage into the fields. In the case of cadmium, it is necessary to admit an error in the analysis of the samples using the neutron activation analysis. It is evident that higher values occur only in the samples collected in the year 2016. Therefore, differences between the results from the years 2015 and 2016 can be counted on.

The differences between the elements from factor 4 and 5 and the elements from other factors are also confirmed by the results of the PCA depicted in the graph of components weights (Figure 6) where Cd, Se, Mn, and Rb show an obvious positive mutual correlation paired with a negative correlation towards Cl and K. In the case of K, intake from soil substrate can occur [56–59]. According to the PCA result, none of these elements correlates with other examined elements. This fact is confirmed by the results of the correlation matrix (Figure 9) where the elements Au, Cl, K, W, Se, Cd, Rb, Cs, Zn, Sb, Mn, Br, and I do not correlate with a greater group of other examined elements. The last element of the group, tungsten, came out to be in the category among the elements meaning contamination in all factors (CF, Igeo, EF) and as such coming from anthropogenic activity. The contamination factor of W belongs to the "moderate" and "severe" categories across the entire studied area except for peripheral areas where protected landscape areas are found. Tungsten is a part of coal and given its high density and difficult melting properties, it is widely used in a number of industrial sectors. For example, it is used in the production of lightbulb threads and W electrodes, the utilization as an ingredient in alloys to increase the hardness and mechanical and heat resistance (fast speed steel), and the production of penetration projectiles or materials for radiation shielding [62]. The contamination factor in the "moderate" category appears almost everywhere across the inhabited area. This fact can be caused by domestic heating where coal burning takes place. Higher concentrations are located in the vicinity of Třinec, Český Těšín and Cieszyna, Ostrava, Rybnik, Rydułtowy, and Wodzisław Śląski where the metallurgical industry is well-developed.

Factor 5 is made of elements Cl and K. The highest values appeared particularly in agricultural areas. Based on the result of Igeo, CF, and EF, potassium is a part of the natural background; it is a biogenic element and therefore, a vegetal origin can be expected. For instance, Olise et al. [42] identified crustal/soil dust as a source. In the case of CF, chlorine is classified into the slight contaminated category. By comparing the distribution of its concentrations with the distribution of the corresponding CF values, locations where high values are a part of the natural background are eliminated (Figure 14).

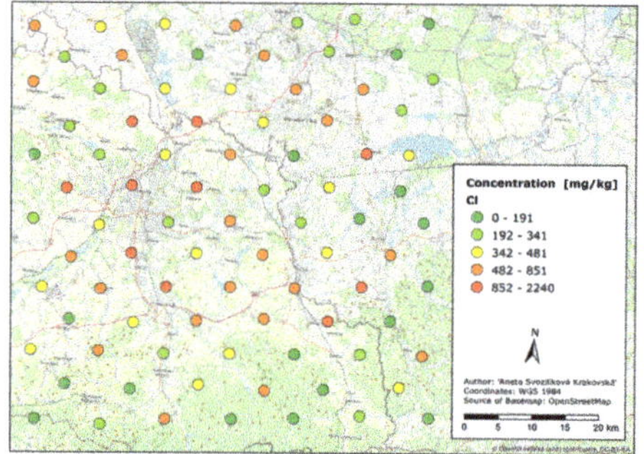

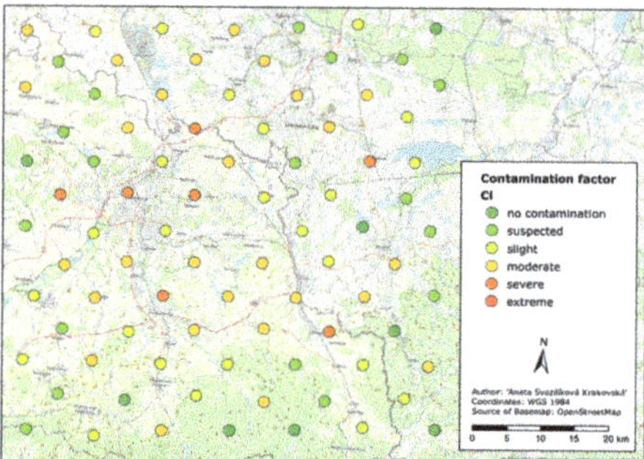

Figure 14. The distribution of concentrations of Cl and the values of the contamination factor for Cl.

Most of the locations with higher CF values are found in the vicinity of agricultural areas and an influence through the use of potassium fertilizers, whose basic component is potassium chloride (KCl) or potassium sulfate (K_2SO_4), can be assumed. Higher values in the areas surrounding Ostrava, Frýdek Místek, and the Polish city Skoczów can originate from chlorine and hydrogen chloride leaks from the industry, specifically from their production and processing, the burning of chloride-containing fuels such as coal, or the leakage of hydrochloric acid during steel processing.

Regarding the rate of excessive of the concentration of heavy metals in the natural environment calculated using the Tomlinson Pollution Load Index (PLI) based on the division of values into categories greater or less than one, expressing if given concentrations in a given location are nearing the value of the background or indicating pollution, the results identified nearly the entire studied area (except for six points in the southern mountainous part) as affected by anthropogenic activity. Similar results were also reached by Shirani et al. [13]. For that reason, a map with PLI values divided into six categories (Figure 11) was created. Higher PLI values correspond with the distribution of industrial centers on both the Czech and the Polish side. High values are also located to the west and south-west of the cities Frýdek-Místek and Paskov, reaching into the Poodří Protected

Landscape Area where higher concentration of industry is not found. This reality can be caused by the north-east drift from industrial areas or domestic heating together with transportation emissions. In this location, an airport can also be found. However, none of the performed statistical analyses determined the link between higher concentrations of individual heavy metals and the vicinity of the airport. Moderately elevated PLI values are observed on the north windy side of the Beskids. Except for typical industrial areas in the studied location, high PLI values are also found in the area surrounding Dolní Benešov where the engineering industry and a foundry are located. A high index also appears in the north part of the Opava region in the Sudice-Třebom promontory. Despite the absence of heavy industry, high concentrations of a number of elements (As, Br, Mn, Zn, Ca, Cr, Fe, Mo, Cd, W, Sc, and Cl) can be found there. This fact can be affected by the surface gypsum mine located in the land registry of the municipality Kobeřice where the subsequent processing of natural gypsum also takes place. Another influence in this area can come from domestic heating, particularly in the town Třebom where according to the results of the 2011 census of persons, houses, and flats [63], heating with solid fuels, particularly with coal, coke, or coal briquettes, prevails. Additionally, transmission of emissions to this area from the Polish side probably takes place.

5. Conclusions

During the years 2015, 2016, 99 samples of moss were collected, the results of which were processed as part of the presented work. After eliminating unsuitable samples, 93 samples were included in the analyses. The samples were analyzed using the instrumental neutron activation analysis. The result was represented by 38 variables, the chemical elements.

A partial goal of this work was to evaluate the measured data and determine if the elements contained in the samples were of natural or anthropogenic origin. To reach this goal, multi-criteria analysis of data was used. Hierarchical clustering on principal components and factor analysis were conducted. Through the principal component analysis, locations most typical for the studied area were determined; these were the locations in the immediate vicinity of metallurgical plants. From the results of this partial goal, it can be assumed that the dominant polluter of the area of interest likely is the metallurgical industry. By hierarchical clustering, the area was divided, according to the type of pollution, into three groups composed of an industry-influenced group, a group with prevailing emissions from domestic heating and also industrial emissions from long-distance transmission, and a so-called clean group.

The distinction between elements with their concentrations near the natural background and elements which can indicate pollution was performed using a number of factors, namely the contamination, geoaccumulation, and enrichment factor and the pollution load index. An intersection of a set of elements Sm, W, U, Tb, and Th, which were determined to be the elements causing pollution and whose concentrations do not near the natural background of the studied locations, took place in all factors. The elements are emitted among others by the metallurgical industry. Sm and Th are applied in a number of chemical industries. Tb is used moreover by electronic industry. W and U are substantial compound of coal. High concentrations in samples coming from areas with no local sources of pollution can be associated with a climactic standpoint, particularly for the prevailing direction of the wind, and long-distance transmission of pollute.

Moss sampling in the study was carried out in an exceptional dense network of sampling sites. Based on the results of the paper, it is possible to propose a dense sampling of the entire region of the Czech Republic as biomonitoring with the use of mosses gives us the ability to detect regional sources of air pollution in a sufficiently dense network yet at acceptable costs.

Author Contributions: Data curation, A.S.K. and V.S.; Formal analysis, A.S.K. and K.V.; Funding acquisition, P.J.; Visualization, V.S.; Writing—original draft, A.S.K.; Writing—review & editing, I.Z. All authors have read and agreed to the published version of the manuscript.

Funding: This research was funded by project solved in cooperation of Czech republic with the JINR (3 + 3 project) "Air pollution characterization in Moravian Silesian region using nuclear and related analytical techniques and GIS Technology", 2015—2017, No. 03-4-1128, main investigators P. Jančík, M. Frontasyeva and was supported in part by Project Interreg Central Europe AIR TRITIA-Uniform Approach to the Air Pollution Management System for Functional Urban Areas in Tritia Region, 2017—2020, No. CE1101, leading partner VSB-Technical University of Ostrava, leading manager P. Jančík.

Conflicts of Interest: The authors declare no conflict of interest.

Abbreviations

The following abbreviations are used in this manuscript:

MDPI	Multidisciplinary Digital Publishing Institute
DOAJ	Directory of open access journals
TLA	Three letter acronym
LD	linear dichroism
i.e.,	id est
EEA	The European Environment Agency
PM	Particulate matter
REZZO	Register of Emissions and Air Polution Sources
GIS	Geographic information system
ICP	International Cooperative Programme
JINR	Joint Institute for Nuclear Research
SLI	short lived isotope
LLI	long lived isotope
HPGe	high pure germanium
NAA	Neutron activation analysis
CRM	Certified reference material
NIST	National Institute of Standards and Technology
SRM	Standard Reference Materials
CF	Contamination factor
Igeo	Geoaccumulation index
PLI	Pollution load index
EF	Enrichment factor
PCA	Principal Component Analysis
HCPC	Hierarchical Clustering on Principal Components
e.g.,	exempli gratia
FA	Factor analysis

References

1. Mulgrew, A.; Williams, P. *Biomonitoring of Air Quality Using Plants*; Air Hygiene Report; WHO Collab. Centre for Air Quality Management and Air Pollution Control: Berlin, Germany, 2000.
2. Satsangi, G.; Yadav, S.; Pipal, A.; Kumbhar, N. Characteristics of trace metals in fine (PM2.5) and inhalable (PM10) particles and its health risk assessment along with in-silico approach in indoor environment of India. *Atmos. Environ.* **2014**, *92*, 384–393. [CrossRef]
3. Sarti, E.; Pasti, L.; Rossi, M.; Ascanelli, M.; Pagnoni, A.; Trombini, M.; Remelli, M. The composition of PM1 and PM2.5 samples, metals and their water soluble fractions in the Bologna area (Italy). *Atmos. Pollut. Res.* **2015**, *6*. [CrossRef]
4. European Environment Information and Observation Network (Eionet). European Environment Agency. Available online: https://www.eea.europa.eu/ (accessed on 26 May 2020).
5. Cherry Hill Mortgage Investment Corp. ISKO: Informace o úrovni Znečištění Ovzduší ve Smyslu Zákona o Ochraně Ovzduší. Available online: http://portal.chmi.cz/ (accessed on 11 March 2019).
6. Cherry Hill Mortgage Investment Corp. Register of Emissions and Air Polution Sources (REZZO). Available online: http://portal.chmi.cz/files/portal/docs/uoco/oez/embil/11embil/rezzo1_4/rezzo1_4_CZ.html (accessed on 26 May 2020).

7. Krakovská, A.S. Analýza Prostorových Dat z Biomonitoringu v Průmyslových Oblastech. Ph.D. Thesis, VSB–Technical University of Ostrava, Ostrava, Czech Republic, 2020.

8. UC Centre for Ecology and Hydrology. ICP Vegetation. Available online: https://icpvegetation.ceh.ac.uk/ (accessed on 26 May 2020).

9. Rodrigues, L.A.; Daunis-i Estadella, P.; Figueras, G.; Henestrosa, S. *Compositional Data Analysis: Theory and Applications*; John Wiley & Sons: Hoboken, NJ, USA, 2011; pp. 235–254. [CrossRef]

10. Eaton, M. *Multivariate Statistics: A Vector Space Approach*; Lecture Notes-Monograph Series; Institute of Mathematical Statistics: Hoboken, New Jersey, USA, 2007.

11. Aitchison, J. *The Statistical Analysis of Compositional Data*; Chapman & Hall, Ltd.: GBR, Australia, 1986.

12. Hron, K. Elementy Statistické Analýzy Kompozičních Dat in Informační Bulletin České Statistické Společnosti, Praha, Czech Republic, 2010, Volume 3. Available online https://www.statspol.cz/oldstat/bulletiny/ib-2010-3.pdf (Accessed on 25 October 2020).

13. Shirani, M.; Afzali, K.; Jahan, S.; Strezov, V.; Soleimani-Sardo, M. Pollution and contamination assessment of heavy metals in the sediments of Jazmurian playa in southeast Iran. *Sci. Rep.* **2020**, *10*. [CrossRef]

14. Jiang, H.H.; Cai, L.; Wen, H.H.; Luo, J. Characterizing pollution and source identification of heavy metals in soils using geochemical baseline and PMF approach. *Sci. Rep.* **2020**, *10*. [CrossRef] [PubMed]

15. Betsou, C.; Tsakiri, E.; Kazakis, N.; Vasilev, A.; Frontasyeva, M.; Ioannidou, A. Atmospheric deposition of trace elements in Greece using moss Hypnum cupressiforme Hedw. as biomonitors. *J. Radioanal. Nucl. Chem.* **2019**, *320*, 597–608. [CrossRef]

16. Koroleva, Y.; Napreenko, M.; Baymuratov, R.; Schefer, R. Bryophytes as a bioindicator for atmospheric deposition in different coastal habitats (a case study in the Russian sector of the Curonian Spit, South-Eastern Baltic). *Int. J. Environ. Stud.* **2020**, *77*, 152–162. [CrossRef]

17. Demková, L.; Árvay, J.; Bobulska, L.; Hauptvogl, M.; Hrstková, M. Open mining pits and heaps of waste material as the source of undesirable substances: Biomonitoring of air and soil pollution in former mining area (Dubnik, Slovakia). *Environ. Sci. Pollut. Res.* **2019**, *26*. [CrossRef] [PubMed]

18. Aničić Urošević, M.; Frontasyeva, M.; Tomasevic, M.; Popovic, A. Assessment of Atmospheric Deposition of Heavy Metals and Other Elements in Belgrade Using the Moss Biomonitoring Technique and Neutron Activation Analysis. *Environ. Monit. Assess.* **2007**, *129*, 207–219. [CrossRef] [PubMed]

19. Vergel, K.; Zinicovscaia, I.; Yushin, N.; Frontasyeva, M. Heavy Metal Atmospheric Deposition Study in Moscow Region, Russia. *Bull. Environ. Contam. Toxicol.* **2019**, *103*. [CrossRef]

20. Thieme, N. R generation. *Significance* **2018**, *15*, 14–19. [CrossRef]

21. Templ, M.; Hron, K.; Filzmoser, P. robCompositions: An R-Package for Robust Statistical Analysis of Compositional Data. In *Compositional Data Analysis: Theory and Applications*; John Wiley and Sons, Ltd.: Chichester, UK, 2011; pp. 341–345.

22. Lê, S.; Josse, J.; Husson, F. FactoMineR: An R Package for Multivariate Analysis. *J. Stat. Softw. Artic.* **2008**, *25*, 1–18. [CrossRef]

23. Wei, T.; Simko, V. *R Package "Corrplot": Visualization of a Correlation Matrix*, Version 0.84. 2017. Available online: https://github.com/taiyun/corrplot/ (accessed on 26 May 2020).

24. StatSoft CR s.r.o. TIBCO Statistica. Available online: http://www.statsoft.cz/ (accessed on 26 May 2020).

25. Barandovski, L.; Frontasyeva, M.; Stafilov, T.; Sajn, R.; Ostrovnaya, T. Multi-element atmospheric deposition in Macedonia studied by the moss biomonitoring technique. *Environ. Sci. Pollut. Res.* **2015**, [CrossRef] [PubMed]

26. Bajraktari, N.; Morina, I.; Demaku, S. Assessing the Presence of Heavy Metals in the Area of Glloogoc (Kosovo) by Using Mosses as a Bioindicator for Heavy Metals. *J. Ecol. Eng.* **2019**, *20*, 135–140. [CrossRef]

27. Zinicovscaia, I.; Aničić Urošević, M.; Vergel, K.; Vieru, E.; Frontasyeva, M.; Povar, I.; Duca, G. Active Moss Biomonitoring of Trace Elements Air Pollution in Chisinau, Republic of Moldova. *Ecol. Chem. Eng.* **2018**, *25*, 361–372. [CrossRef]

28. Shetekauri, S.; Chaligava, O.; Shetekauri, T.; Kvlividze, A.; Kalabegishvili, T.; Kirkesali, E.; Frontasyeva, M.; Chepurchenko, O.; Tselmovich, V. Biomonitoring Air Pollution Using Moss in Georgia. *Pol. J. Environ. Stud.* **2018**, *27*. [CrossRef]

29. Korzekwa, S.; Frontasyeva, M. Air Pollution Studies in Opole Region, Poland, Using the Moss Biomonitoring and INAA. *Am. Inst. Phys.* **2007**, *958*, 230–231. [CrossRef]

30. Ermakova, E.; Frontasyeva, M.; Pavlov, S.; Povtoreiko, E.; Steinnes, E.; Cheremisina, Y. Air Pollution Studies in Central Russia (Tver and Yaroslavl Regions) Using the Moss Biomonitoring Technique and Neutron Activation Analysis. *J. Atmos. Chem.* **2004**, *49*, 549–561. [CrossRef]

31. Haruštiaková, D.; Jarkovský, J.; Littnerová, S.; Dušek, L. *Vícerozměrné Statistické Metody v Biologii*; Akademické nakladatelství CERM, s.r.o.: Brno, Czech Republic, 2012.

32. Hebák, P. *Vícerozměrné Statistické Metody*; Informatorium: Praha, Czech Republic, 2007.

33. Legendre, P.; Legendre, L. *Numerical Ecology*, 3rd ed.; Elsevier: Amsterdam, The Netherlands, 2012.

34. Håkanson, L. An Ecological Risk Index for Aquatic Pollution Control—A Sedimentological Approach. *Water Res.* **1980**, *14*, 975–1001. [CrossRef]

35. Fernández, J.; Rey, A.; Carballeira, A. An extended study of heavy metal deposition in Galicia (NW Spain) based on moss analysis. *Sci. Total Environ.* **2000**, *254*, 31–44. [CrossRef]

36. Muller, G. Index of Geoaccumulation in Sediments of the Rhine River. *Geojournal* **1969**, *2*, 109–118.

37. Turekian, K.; Wedepohl, K. Distribution of the Elements in Some Major Units of the Earth's Crust. *Geol. Soc. Am. Bull.* **1961**, *72*. [CrossRef]

38. Covelli, S.; Fontolan, G. Application of a normalization procedure in determining regional geochemical baselines. *Environ. Geol.* **1997**, *30*, 34–45. [CrossRef]

39. Sinex, S.; Helz, G. Regional geochemistry of trace elements in Chesapeake Bay sediments. *Environ. Geol.* **1981**, *3*, 315–323. [CrossRef]

40. Rubio, B.; Nombela, M.; Vilas, F. Geochemistry of Major and Trace Elements in Sediments of the Ria de Vigo (NW Spain): An Assessment of Metal Pollution. *Mar. Pollut. Bull.* **2000**, *40*, 968–980. [CrossRef]

41. Ackermann, F. A procedure for correcting the grain size effect in heavy metal analyses of estuarine and coastal sediments. *Environ. Technol. Lett.* **2008**, *1*. [CrossRef]

42. Olise, F.; Lasun Tunde, O.; Olajire, M.; Owoade, O.; Oloyede, F.; Fawole, O.G.; Ezeh, G. Biomonitoring of environmental pollution in the vicinity of iron and steel smelters in southwestern Nigeria using transplanted lichens and mosses. *Environ. Monit. Assess.* **2019**, *191*. [CrossRef]

43. Rotter, P.; Šrámek, V.; Vácha, R.; Borůvka, L.; Fadrhonsová, V.; Sáňka, M.; Drábek, O.; Vortelová, L. Rizikové prvky v lesních půdách: Review. *ZpráVy Lesn. VýZkumu* **2013**, *58*, 17–27.

44. Kłos, A.; Rajfur, M.; Wacławek, M.; Wacławek, M. Application of enrichment factor (EF) to the interpretation of results from the biomonitoring studies. *Ecol. Chem. Eng.* **2011**, *18*, 171–183.

45. Tomlinson, D.; Wilson, J.; Harris, C.; Jeffrey, D. Problems in Assessment of Heavy Metals in Estuaries and the Formation of Pollution Index. *HelgoläNder Meeresunters.* **1980**, *33*, 566–575. [CrossRef]

46. Salo, H.; Bućko, M.; Vaahtovuo, E.; Limo, J.; Mäkinen, J.; Pesonen, L. Biomonitoring of air pollution in SW Finland by magnetic and chemical measurements of moss bags and lichens. *J. Geochem. Explor.* **2012**, *115*, 69–81. [CrossRef]

47. Angulo, E. The Tomlinson Pollution Load Index applied to heavy metal, 'Mussel-Watch' data: A useful index to assess coastal pollution. *Sci. Total Environ.* **1996**, *187*, 19–56. [CrossRef]

48. Aitchison, J.; Greenacre, M. Biplots of Compositional Data. *J. R. Stat. Soc. Ser.* **2002**, *51*, 375–392. [CrossRef]

49. Cartell, R. The Scree Test for the Number of Factors. *Multivar. Behav. Res.* **1966**, *1*, 245–276. [CrossRef] [PubMed]

50. Meloun, M. *Vícerozměrná Analýza Dat Metodou Hlavních Komponent a Shluků*; Analýza Dat 2015; Celostátní Seminář a Workshop: Pardubice, Czech Republic, 2015; pp. 25–42.

51. Barbieri, M. The Importance of Enrichment Factor (EF) and Geoaccumulation Index (Igeo) to Evaluate the Soil Contamination. *J. Geol. Geophys.* **2016**, *5*, [CrossRef]

52. Nowrouzi, M.; Pourkhabbaz, A. Application of geoaccumulation index and enrichment factor for assessing metal contamination in the sediments of Hara Biosphere Reserve, Iran. *Chem. Speciat. Bioavailab.* **2014**, *26*. [CrossRef]

53. Pant, P.; Harrison, R. Critical Review of Receptor Modelling for Particulate Matter: A Case Study of India. *Atmos. Environ.* **2012**, *49*, 1–12. [CrossRef]

54. Fawole, O.; Owoade, O.; Hopke, P.; Olise, F.; Lasun Tunde, O.; Olaniyi, B.; Jegede, O.; Ayoola, M.; Bashiru, M. Chemical compositions and source identification of particulate matter (PM2.5 and PM2.5–10) from a scrap iron and steel smelting industry along the Ife–Ibadan highway, Nigeria. *Atmos. Pollut. Res.* **2015**, *6*, 107–119. [CrossRef]

55. Kara, M.; Hopke, P.; Dumanoglu, Y.; Altiok, H.; Elbir, T.; Odabasi, M.; Bayram, A. Characterization of PM Using Multiple Site Data in a Heavily Industrialized Region of Turkey. *Aerosol Air Qual. Res.* **2015**, *15*, 11–27. [CrossRef]

56. Bates, J.; Farmer, A. *Bryophytes and Lichens in a Changing Environment*; Oxford University Press: New York, NY, USA, 1992.

57. Wells, J.; Boddy, L. Phosphorous translocation by saprotrophic basidiomycete cord systems on the floor of a mixed deciduous woodland. *Mycol. Res.* **1995**, *99*, 977–980. [CrossRef]

58. Brown, D.; Brumelis, G. A biomonitoring method using the cellular distribution of metals in moss. *Sci. Total Environ.* **1996**, *187*, 153–161. [CrossRef]

59. Brumelis, G.; Brown, D. Movement of Metals to New Growing Tissue in the Moss Hylocomium splendens (Hedw.) BSG. *Ann. Bot.* **1997**, *79*. [CrossRef]

60. Pandey, V.; Upreti, D.; Pathak, R.; Pal, A. Heavy Metal Accumulation in Lichens from the Hetauda Industrial Area Narayani Zone Makwanpur District, Nepal. *Environ. Monit. Assess.* **2002**, *73*, 221–228. [CrossRef] [PubMed]

61. Praha: Ministerstvo životního Prostředí, CENIA. Integrovaný Registr Znečišťování životního Prostředí. Available online: https://www.irz.cz/ (accessed on 26 May 2020).

62. Greenwood, N.N.; Earnshaw, A. *Chemie Prvků*; Informatorium: Praha, Czech Republic, 1993.

63. Český Statistický úřad (ČSÚ). Český Statistický úřad. Available online: https://www.czso.cz/ (accessed on 26 May 2020).

MDPI

St. Alban-Anlage 66

4052 Basel

Switzerland

Tel. +41 61 683 77 34

Fax +41 61 302 89 18

www.mdpi.com

Atmosphere Editorial Office

E-mail: atmosphere@mdpi.com

www.mdpi.com/journal/atmosphere